Contents

Acknowledgements

My sincere appreciation is extended to Armin Freitag, Rob Lowden, Mark Twiname, Bertam Nold, Sébastien Haule and Rico Unger for generously sharing their knowledge and expertise in making this book possible. I would also like to thank Christian Ankerstjerne (webmaster of Panzerworld.com) and Akira Takiguchi for generously sharing images from their personal collection, Rico Unger, co-author of *Panzerregiment 1: From origin to the Polish Campaign* (Designfeld), and Peter Samsonov. I would especially like to acknowledge and thank Hilary Louis Doyle, and the late Thomas L. Jentz, for their *Panzer Tracts* publications – indispensable references for any student of German Second World War armour. Finally, my thanks to Sarah Cook for editing the manuscript, Noel Sadler for his design expertise in bringing the book to life and Michael Leventhal from Greenhill Books for championing the importance of history.

Abbreviations

Ausf.	*Ausführung*
AWM	Australian War Memorial
hp	horsepower
kl.Pz.Bef.Wg	*kleiner Panzerbefehlswagen*
km/h	kilometres per hour
La.S	*Landwirtschaftlicher Schlepper*
mph	miles per hour
Pz.Kpfw.	*Panzerkampfwagen*
NARA	US National Archives and Records Administration
Sd.Kfz.	*Sonderkraftfahrzeug* (special motor vehicle)
KwK	*Kampfwagenkanone*
NCO	Non-commissioned officer
S.m.K.	*Spitzgeschoss mit Kern*

IMAGES OF WAR

PANZERTRUPPE PANZER I 1934–1941

RARE PHOTOGRAPHS FROM WARTIME ARCHIVES

David Mitchelhill-Green

Greenhill Books

First published in Great Britain in 2026 by

Greenhill Books

c/o Pen & Sword Books Ltd, 47 Church Street, Barnsley,
South Yorkshire, S70 2AS, England
For more information on our books, please visit
www.greenhillbooks.com, email contact@greenhillbooks.com
or write to us at the above address.

ISBN 978-1-80500-244-4
ePub ISBN 978-1-80500-245-1
PDF ISBN 978-1-80500-246-8

A CIP catalogue record for this book is available from the British Library.

Typeset in Gill Sans by Concept, Huddersfield, West Yorkshire, HD4 5JL
Printed and bound in England by CPI Group (UK) Ltd, Croydon, CR0 4YY

Introduction

A wartime book in Nazi Germany entitled *Schnelle Truppen* ('Mobile Troops') narrated the history of the *Panzerkampfwagen* (lit. 'armoured fighting vehicle'), beginning in 1918. With the end of the [First World] war, it described how:

[t]he development of armoured combat vehicles shifted from combat on the field to the test site. All nations, including the then completely disarmed Germany, worked on the further development of tanks. In Germany, this could only be done in secret and mostly on the drawing board. The construction of *Panzerkampfwagen* made enormous progress. The size and target area became smaller, speed and manoeuvrability grew. Today, caterpillar tracks allow the tanks to drive on the road with a service life of many thousands of kilometres. Their rotating turrets carry armour-piercing guns and machine guns . . . Thus, the cumbersome tools of trench warfare have become a high-quality instrument of mobile warfare, armoured weapons that can be quickly driven at the enemy.

This is the story of German interwar tank development, culminating in Adolf Hitler's first mass-produced tank – the *Panzerkampfwagen* I, including its variants and combat service.

Chapter One

The Kaiser's Panzers and Interwar Development

The concept of attacking an enemy with a motorized armoured vehicle came of age on 15 September 1916 with the introduction of Britain's 'tank' at the Battle of Flers-Courcelette. Although failing to satisfy Whitehall's expectations of a decisive victory, the first Mark I tank nevertheless demonstrated its potential for tactical support – drawing fire from accompanying infantry, lessening the machine-gun's battlefield dominance, overwhelming enemy strong-points and demoralizing enemy infantry. A modicum of mobility had returned to the Western Front, with invaluable lessons learnt, including the need for careful planning and reconnaissance, exhaustive training and adequate communication. Improved mechanical reliability, increased speed and thicker armour would feature on subsequent models.

This British Mark II male tank, No. 799, was abandoned at Bullecourt following a failed attack on 9 April 1917. Crucially, the armour of this training tank was not subjected to the final hardening process during manufacture, leading the Germans to believe that the machine posed only a limited threat, one that could be easily countered using existing weapons with armour-piercing 7.92mm S.m.K. (*Spitzgeschoss mit Kern*, 'Kern' referring to an iron core) ammunition and tactics.

Sturmpanzerwagen (armoured assault vehicle) A7V *Wotan*. The A7V was Imperial Germany's only operational tank. Designed by pioneering tank engineer Joseph Vollmer, the vehicle was armed with a Cockerill-Nordenfelt rapid-fire 5.7cm L/26.3 cannon and six MG-08 machine guns. Crewed by eighteen men, the 30-ton tank had an on-road speed of up to 16km/h and an off-road speed of 4–8km/h. Comparing the A7V with foreign tanks, Ernst Volckheim, A7V commander and future armoured warfare theorist, wrote: 'In contrast to the layout of the enemy tanks the tracks of the A7V-*Sturmpanzerwagen* were protected by armour plates. The fact that each track was powered by one engine resulted in a better steering performance. Due to the thicker armour, vulnerability to enemy fire was lower. On the other hand, the mobility of the A7V in rugged terrain, under constant artillery fire, was significantly worse. Trenches and shell craters caused serious problems.' Only twenty examples were built, with some local tactical success achieved during the final year of the war.

Although the first Allied tanks failed to impress many senior officers within Germany's Supreme Army Command, a meeting of the so-called 'A7V Committee' on 30 October 1916 brought together experts from some of the country's leading automotive firms to discuss building a German equivalent to the tank. Plans for the prototype vehicle were unveiled on 22 December 1916, based on a lengthened version of a Holt tractor chassis and powered by twin Daimler 100hp engines. Just weeks later Daimler-Motoren-Gesellschaft (DMG) in Berlin-Marienfelde unveiled a mock-up of the hull with the vehicle's first demonstration held at the company's proving ground on 30 April 1917.

(**Opposite, above**) Turncoat British Mark IV female. Imperial Germany addressed its shortcomings in armour by reconditioning dozens of captured tanks, or *Beutepanzer*, at Charleroi in occupied Belgium. Plans to manufacture a direct copy of the British vehicle were shelved. The Mark IV, in Volckheim's opinion, was an inferior vehicle: 'it had a clearly weaker fighting power than the A7V. Due to the exposed tracks and the weaker armour protection, the tank was vulnerable to infantry fire. Driver and commander were placed in the front of the vehicle. While the driver operated the engine and the transmission, the commander had to handle the steering brakes to initiate turns …'

(**Below**) In early 1918 Vollmer designed a light tank known as the *Leichter Kampfpanzer*, or LK I. It was based on an existing Daimler car chassis, with drive sprockets replacing the rear wheels. Weighing 7 tons, the machine featured 8mm of armour and a single 7.92mm MG-08 machine gun in a revolving turret. The front-mounted Daimler-Benz four-cylinder, 60hp engine delivered a top speed of 12mph (20km/h).

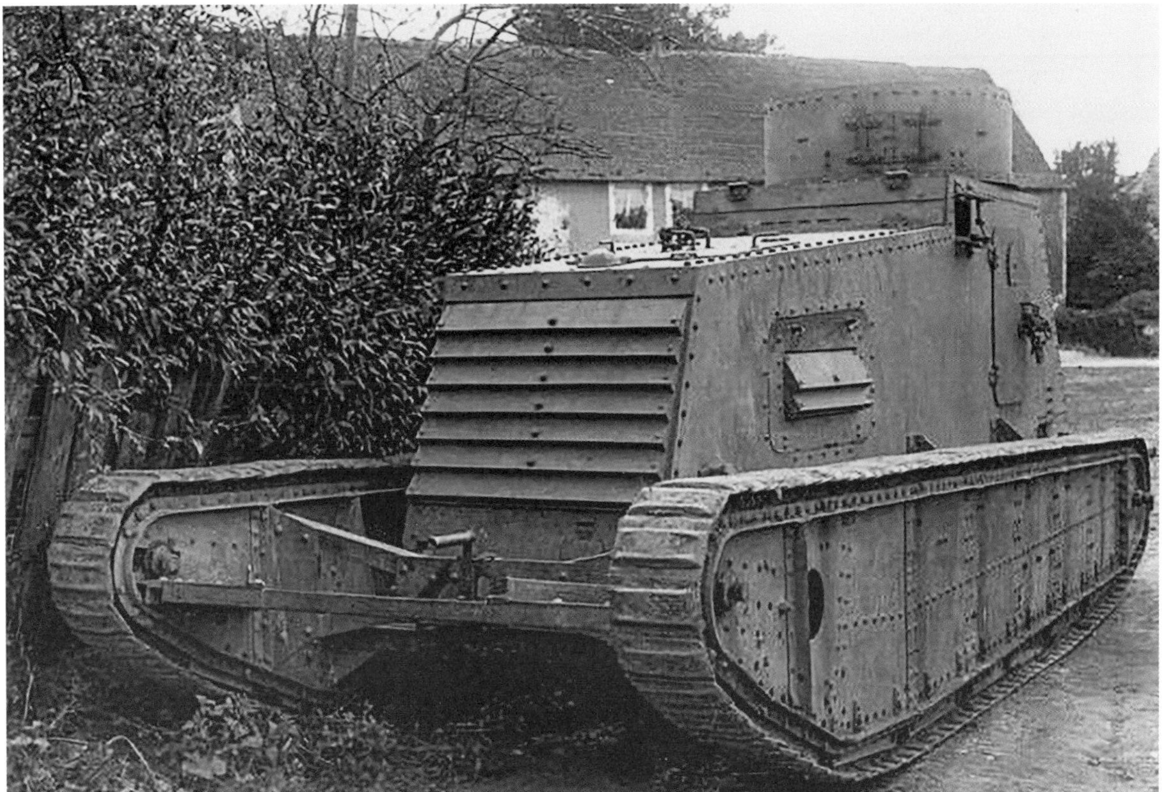

The heavier *Leichter Kampfwagen* II (LK II) was protected by up to 14mm of armour and was equipped with a 37mm gun. Also powered by a Daimler-Benz 60hp engine, the vehicle achieved a speed of 7mph (12km/h) during trials. Vollmer also began work on a third model, the *Leichter Kampfwagen* III (LK III), with an internal layout like that of France's Renault FT, before the war ended, but none was built. Two further prototype designs for another light tank, the 19-ton *Sturm-panzerwagen Oberschlesien*, were partially completed by October 1918.

Interestingly, a representation of the *Leichter Kampfwagen* II appeared on a May 1922 postcard celebrating German motorized troops.

While German tank development continued, including the colossal 120-ton *Großkampfwagen*, the tide of war had turned for Imperial Germany. General Erich Ludendorff considered 8 August 1918, the beginning of the Allied Hundred Days Offensive, as a 'black day for the German Army'. The massive Allied Western Front onslaught – a foretaste of Second World War mobile warfare – involved coordinated assaults by British, Commonwealth, French and American infantry, armour and aircraft, which shattered the depleted and exhausted German Army, pushing it back to the battlefield of 1914. Seeking peace terms, Ludendorff's representative, Major Erich von dem Bussche-Ippenburg explained to politicians inside the Reichstag on 2 October 1918 the pivotal role that the tank was playing:

> The enemy has made use of tanks in unexpectedly large numbers. In cases where they have suddenly emerged in huge masses from smoke clouds, our men were completely unnerved. Tanks broke through our foremost lines, making a way for their infantry, reaching our rear, and causing local panics, which entirely upset our battle control. When we were able to locate them our anti-tank guns and our artillery speedily put an end to them. But the mischief had already been done, and solely owing to the success of the tanks we have suffered enormous losses in prisoners …

Fighting continued until the Armistice took effect on 11 November 1918. Extended three times, the voluminous treaty formally ending the First World War was signed at Versailles on 28 June 1919. Ratified and put into effect on 10 January 1920, the controversial agreement ascribed responsibility for the war to Germany and other Central Powers, and sowed the seeds for future aggression. Berlin was forced to cede territory to France, Belgium and Poland, to pay reparations and to surrender its overseas colonies. With Britain and France fearful of communism taking hold, Article 160 dictated a Germany Army not exceeding 100,000 men and not containing more than seven infantry and three cavalry divisions. Heavy artillery, poisonous gas and military aircraft were banned, while Article 171 prohibited 'the manufacture and the importation into Germany of armoured cars, tanks and all similar constructions suitable for use in war'. Germany, therefore, would possess an emasculated army, a home defence force incapable of attacking its neighbours.

In accordance with the peace treaty, the Weimar Republic s fledgling military turned to exercises using mock-up tanks while covertly embarking upon a pro-gramme to export the LKII and develop a series of 'agricultural tractors'.

Leichter *Kampfwagen* II in Sweden and Hungary

'Export tractor' – basically an LK II chassis sans superstructure, Vollmer's design saw postwar service in Sweden and Hungary. Obtaining quotes from Germany for *Raupenschlepper* ('tracked tractors'), Sweden purchased ten tanks. Concealed from prying Allied eyes, parts were shipped from Lübeck in June 1922 under a customs declaration stating the shipment comprised agricultural tractors and steam engine sheet metal. An engineer from Wilhelm Ugè GmbH in Berlin oversaw the assembly of the first vehicle by Stockholm Tygstation. The remaining nine LK II were completed at the naval dockyards in Stockholm.

In Swedish service the former German tank was known as the *Stridsvagn* m/21 (Strv m/21). German general and armour theoretician Heinz Guderian spent four weeks in Sweden in 1929 observing the tanks in army exercises, even driving one himself. Suitably impressed, he recalled …

… 'With their armour plating, their tracks, their guns and their machine guns, they succeeded in carrying their crews, alive and capable of fighting, through artillery barrages and wire entanglements, over trench systems and shell craters, into the centre of German lines. The power of the offensive had come back into its own.'

Strv m/21 '8'. Five vehicles were reconditioned in 1929 as the *Stridsvagn* m21-29 with the more powerful Scania-Vabis 85hp engine. Joseph Vollmer's 1918 design soldiered on in Sweden until 1938, when it was replaced by the Czech CKD AH-IV tankette, known as the Strv m/37. By request, a retired Strv m/21 was reportedly sent to Germany as a museum piece, only to disappear during the Second World War.

German rearmament

The foundations of the fledgling Weimar Republic's *Reichswehr* ('Reich Defence') were laid by the Freikorps, remnants of the wartime Imperial Army that assisted in suppressing revolutionaries and restoring order. *Generaloberst* Johannes Friedrich 'Hans' von Seeckt (1866–1936), First World War hero and recipient of the *Pour Le Merite*, was head of the *Truppenamt* (lit. 'Personnel Office'), as the *Reichswehr*'s veiled General Staff was known. Instrumental in restoring confidence in the army following the divisive March 1920 *Kapp Putsch*, Seeckt set out to create a (treaty-restricted) *Neuzeitliches Heer* or 'modern army'. Viewing mobility as the key to future battlefield success, Seeckt envisaged a small and highly mobile army working in close cooperation with aircraft. Formulating an innovative military doctrine, he turned to covertly developing new weaponry, undermining the terms of the Treaty of Versailles through his focus on technology.

(**Opposite, above**) Hungary's LK II-based *vontatók* ('tractor'). Under the terms of the 1920 Treaty of Trianon, Hungary's army was reduced to 35,000 men. Like Germany, Hungary was banned from manufacturing or importing tanks. Nevertheless, during the 1920s the Hungarian Army clandestinely obtained fourteen LK II tanks, likely in parts, under the guise of importing agricultural tractors. Secretly shuffled from location to location to evade Entente detection, five tanks and one training vehicle eventually became operational. When they proved obsolete and unreliable, Hungary turned to Italy and purchased five *carro d'assalto* Fiat 3000B tanks, an equally outdated design based on the French Renault FT.

(**Opposite, below**) After the war Joseph Vollmer moved to Czechoslovakia, where he joined Škoda. His wheel-cum-track light tank, known as the KH-50 (KH was an abbreviation of *Kolohousenka*, from the Czech words *kolo* and *housenka* meaning 'wheel' and 'caterpillar') was produced in 1923–25; this was followed by the KH-60 in 1928 and KH-70 in 1930. An improvement on the KH-50, Vollmer's prototype KH-60 (pictured here) was abandoned in favour of the British Carden-Loyd tankette. It was discovered by German troops in 1939 and scrapped.

(**Below**) Two former British Mark IV tanks briefly saw postwar service with the Freikorps against communist revolutionaries attempting to overthrow the Weimar government.

With covert rearmament under way, propaganda circulated in the Weimar Republic showed the demise of a former *Beutepanzer*. The image was captioned: 'Disarmament is the first perquisite for solving the world crisis! The work of the Disarmament Conference does not seem to lead to any goal, as France is not willing to base its own disarmament on the disarmament provisions of the Versailles Treaty, which Germany has fully implemented.'

With the Treaty of Versailles prohibiting Germany from producing and possessing tanks, the *Reichswehr* turned to training with mock-up armour. (Galland Books)

Paper panzers. Unimpressed, *General der Kavallerie* Maximillan von Poseck opposed the idea of tanks, even arguing on occasion that his cavalry was better suited to battle because of its faster deployment. (Galland Books)

The Dixi 3/15 PS DA1 was the basis for a multi-piece, mock-up tank. In 1927 Dixi-Werke AG began producing a licence-built copy of Britain's Austin 7 known as the Dixi 3/15 PS DA-1. The DA-1 designation referred to the *Erste Deutsche Ausführung* (first German version). Taken over by Bayerische Motoren Werke AG (BMW) the following year, the upgraded car became known as the BMW Dixi 3/15 PS DA-2 or simply the BMW 3/15 PS DA-2. Two further versions were produced until the licence with Austin expired and BMW produced its 3/20 PS model. A makeshift lightly armoured version of the BMW 3/15 PS proved unsuccessful.

RW-3249

Reichswehr Panzerattrappen (dummy tanks) on winter manoeuvres. Early lessons learned included the advantage of surprise and the use of 'armour' in waves. One of the first exercises, under *General-leutnant* Otto von Stülpnagel, was held in Grafenwöhr in May 1928. *Inspekteur der Verkehrstruppen* (Inspector of Transport Troops) Alfred von Vollard-Bockelberg divided the 'tanks' into three waves – the first two neutralized 'enemy' artillery and machine-gun positions, while the third worked closely with advancing infantry – mirroring British tactics honed from wartime experience. At the same time an evaluation of a possible Polish or French attack concluded that 'any military engagement involves the danger of catastrophe', with Germany unable to defend itself against either aggressor.

In recognition of Germany's original *Panzertruppen*, Otto Gessler, the Minister of Defence, instigated the *Kampfwagenabzeichen* (tank badge) on 13 July 1921. The award, which featured a *Sturm-panzerwagen* A7V on a battlefield with shrapnel shells bursting overhead, was available privately to former crewmen of either an A7V or *Beutepanzer* who had participated in three assaults or had been wounded in action.

Kama – Germany's secret test facility

With Sweden wary of the *Reichswehr*'s experimentation with armour on its soil, Seeckt turned to the Soviet Union. With the Treaty of Rappallo (signed on 16 April 1922) normalizing economic and diplomatic relations between the future aggressors, a secret addendum four months later set in motion covert military cooperation. Negotiations in October 1926 led to the *Reichswehr* establishing a tank school near Kazan, some 800km (500 miles) east of Moscow, on the site of a former artillery barracks and firing range. It was agreed that Germany would cover the construction costs, while the Soviets would undertake construction and repairs. The facility was codenamed 'Kama' and was later renamed TEKO (Technical Courses of the Society for Defence, Aviation and Construction of Chemical Weapons). German and Russian armour students – the nucleus of future teaching staff in both armies – were trained at Kama between 1929 and 1933. Walter Model and Wilhelm Ritter von Thoma were two German participants who later became senior Wehrmacht commanders.

Philipp Scheidemann, leader of Germany's Social Democrats, later leaked details of the clandestine cooperation inside the Reichstag, declaring a motion of no confidence in the government. With newspapers in the West declaring a 'Crisis in Germany', and apprehension within the Soviet and German militaries, progress at Kazan moved slowly, with Seeckt and his staff concealing all research from Reichstag politicians.

Rheinmetall's experimental 7.7cm *WD Schlepper* 50 PS was a self-propelled gun with a modified 7.7cm FK 96/16 gun mounted on a Hanomag *WD Schlepper* 50 PS. Likely crewed by two men, the underpowered vehicle offered only a low speed and poor cross-country performance. A later vehicle featuring a 7.5cm gun was the forerunner of the *Sturmgeschütz* series.

Reichswehr armoured vehicles tested at Kama included two modified commercially available Hannoversche Maschinenbau AG (Hanomag) agricultural tractors (the 3.7cm *Tak auf WD Schlepper* 25 PS and the 7.7cm *WD Schlepper* 50 PS), the *Großtraktor*, the *Leichttraktor* and the *Rader-Raupen Kampfwagen* M-28. The experience and knowledge gained with these experimental vehicles would contribute to the development of new tanks such as the *Neubaufahrzeug* (new construction vehicle, Nb.Fz.) and the mass-produced *Panzerkampfwagen* (Pz.Kpfw.) – a series of progressively heavier tanks that would spearhead German army actions during the opening campaigns of the Second World War.

Großtraktor

The *Großtraktor* (large tractor) – the codename given to an experimental medium tank jointly developed by Rheinmetall, Daimler-Benz and Krupp – was the first German tank to be tested at Kama. Each company shipped two soft-steel prototypes for testing. Detailed technical requirements issued in 1926 for the development of an *Armeewagen* 20 specified a 6m-long tank, weighing 16 tons. With a crew of six, including a radio operator, the tank was to be armed with a 7.5cm gun in a fully traversable turret, plus three machine guns (including one in a small secondary turret). An existing aircraft engine would deliver a maximum speed of 40km/h, while 14mm of armour would provide protection against steel-core small-arms fire. Two rear propellers would propel the tank in water.

A completed Daimler-Benz *Großtraktor* at Unterlüß prior to shipping to the Soviet Union. Former *Oberleutnant* Klaus Müller recalled that the tank's experimental tracks with rubber pads 'encountered too much resistance' when steering. He added: 'The greased track pins [also] did not work out at all, since water and sand entered through the pin gaskets, thus providing an extremely effective abrasive leading to premature wear. The desired larger roadwheels could not be mounted on these tractors.'

Both the Rheinmetall and Daimler machines were assembled at the Rheinmetall works in Unterlüß, Lower Saxony; the Krupp examples were built at the company's Meppen firing range, also in Lower Saxony. All the vehicles were covertly shipped to Kama via Leningrad in late June 1929. Engineers would grapple with suspension and transmission issues during the subsequent testing, the Rheinmetall model being the worst performing of the three designs.

A Krupp *Großtraktor* prepares to traverse an anti-tank trench. The tank, 6.42m in length, weighed 16.4 tons. Like the two competing designs, it had a fording depth of 100cm, a maximum speed of around 44km/h and a range of 150km. The Krupp and Rheinmetall prototypes featured a BMW Va six-cylinder aircraft engine, delivering 320hp at 1,565rpm, whereas the Daimler-Benz model used a Daimler Type D IVb six-cylinder aircraft engine, producing 260hp at 1,450rpm. Although the tanks were outwardly similar, the three companies used different suspension systems: the Rheinmetall model used hydraulic suspension, Krupp used coil springs, and the Daimler-Benz featured leaf springs. A report signed by *General der Panzertruppe* Oswald Lutz concluded that a Panzer division was purely offensive, whether in attack or in defence. Two roles were advocated: independent strikes against an enemy's rear and flank, or cooperation with other formations. Military cooperation with the Soviets had ceased two years earlier after Adolf Hitler came to power. The *Großtraktor* and *Leichttraktor* vehicles were shipped back to Germany, arriving in Stettin by 21 September 1933. From here the *Großtraktoren* were freighted to the Daimler-Benz facility at Berlin-Marienfelde. The Daimler models were withdrawn from service, becoming monuments at Erfurt and Wünsdorf. Training with the remaining four machines continued.

The old *Versuchserie* (trial series) tanks were relegated to a role as static monuments in various *Panzertruppen Kaserne* (barracks). This Krupp *Großtraktor*, in three-tone camouflage, was installed at the *Schießschule der Panzertruppen* (armoured troops gunnery school) in Germany's Schleswig-Holstein region. The area remains a Bundeswehr training facility.

Krupp *Großtraktor* '43' at Putlos with the later memorial to Oberst Herbert Baumgart, commander of the 2nd Battalion, 3rd Panzer Regiment and former commander of the *Schießschule der Panzertruppen* (1935–August 1938). Baumgart died on the night of 7/8 September 1939 in Biskupice, southern Poland, when his Panzer was hit by Polish artillery fire.

Rear view of Rheinmetall-Borsig *Großtraktor* '46' showing the auxiliary rear turret with a single 7.92mm machine gun.

One of the problematic Daimler-Benz *Großtraktoren*, '41' , was erected as a memorial in front of the 1st Panzer Regiment's Löberfeld barracks at Erfurt, in the central German state of Thuringia.

A Daimler-Benz *Großtraktor* (with fake gun barrel) was the centrepiece of a 'fighting vehicle memorial' near the entrance to the 5th Panzer Regiment's General Lutz barracks in Wünsdorf-Zossen. It was dedicated on 16 March 1936, and the commander of the 2nd Battalion, *Oberstleutnant* Hermann Breith, commented: 'It is intended as a symbol of our unshakable martial spirit and as an incentive to the work in forming the battalion.' Note the empty machine gun aperture on the front hull and the protected central headlight (below).

Leichttraktor (Vs.Kfz.31)

Specifications were issued in mid-1928 by the *Waffenamt* (WaA: the Weimar and Third Reich weapons agency) for the development of a 6-ton experimental *Leichttraktor* ('light tractor') armed with a 3.7cm L/45 gun in a fully rotating turret. A front-mounted Daimler M36 six-cylinder 100hp engine would deliver a road speed of 25–30kph over an operational range of 150–200km. Krupp and Rheinmetall each produced two prototypes in April–May 1930.

Front and side views of Krupp's *Leichttraktor*. The crew of four comprised a driver and radio operator in hull positions, and the commander (also loader) and gunner in the turret.

Krupp's *Leichttraktor* featured leaf spring suspension with six pairs of road wheels and a single leaf spring, and a separate road wheel before the front idler, with two return rollers and a rear drive sprocket. The Rheinmetall machine, conversely, featured coil springs.

A Krupp *Leichttraktor* with modified coil spring suspension, with nine road wheels and the addition of mudguards and a revised front air intake grille. Note also the *Gefechtsantenne* (frame antenna). (Galland Books)

Side and three-quarter views of the Rheinmetall *Leichttraktor*, in three-tone camouflage, with the 3.7cm L/45 Pak gun in a smaller turret. The crew was reduced to three, with the commander also acting as gunner. Note the large turret hatch and rear doors. Following testing at Kama, all four *Leichttraktoren* were returned to Germany.

A Rheinmetall *Leichttraktor* was later modified with a Christie-style suspension featuring four large road wheels.

Another vehicle tested at Kama was the unconventional *Rader-Raupen Kampfwagen* M-28 ('wheels-tracks-fighting vehicle' model 1928), or *Landsverk* L-5 as it was known in Sweden. It was produced by AB Landsverk to a design by GHH engineer Otto Merker. It is seen here in wheeled operation, with the rear wheels driven by chains. Its armament consisted of a 3.7cm cannon and two 7.92mm machine guns. Inside the compact vehicle was a crew of four: commander, gunner, and forward and reverse drivers. Six examples were built. While the Germans were disappointed with the L-5's performance, the wheel-cum-track concept featured in the later Swedish Landsverk L-30.

Reichswehr memorial to the 2,400 German vehicle drivers who died in the First World War, pictured on the cover of *Die Woche* ('The Week'). Created by Fritz Ebhardt and inaugurated on 7 June 1931, the memorial is topped with a fantasy tank with a domed turret resembling the proposed *Sturmpanzerwagen Oberschlesien*.

Seen here after its transfer to the city of Potsdam during the Third Reich, Ebhardt's memorial was destroyed after 1945.

A curious participant in the 1934 Munich carnival was a fictitious tank resembling a British Mark IV. The photo caption reads: 'The Angel of Peace sits enthroned high on the tank, protected from all who would approach him as a foe, but also inaccessible to those who desire his rule.'

Neubaufahrzeug

After examining the small and large 'tractors' at Kama, *General* Heinz Guderian expressed interest in a new *mittlerer Traktor* (medium tractor). Specifications were issued for a multi-turreted vehicle weighing 15 tons. It was originally to be armed with a 7.5cm gun, but a subsequent directive in February 1933 specified a twin turret mounting for a 7.5cm gun and a 3.7cm gun. Both Krupp and Rheinmetall-Borsig submitted designs for the *Panzerkampfwagen Neubaufahrzeug* (Nb.Fz.) (literally, 'new construction vehicle') after October 1933, and the latter's design proved superior. Five tanks were produced: the first vehicle had a Rheinmetall chassis and turret; the second featured a Rheinmetall chassis and Krupp turret, a combination carried forward in the following three production vehicles.

(**Below**) Nb.Fz. under construction. Like the *Großtraktor*, it was fitted with a BMW Va aircraft engine. The running gear on each side comprised a rear drive sprocket, ten coil-sprung road wheels arranged in pairs, plus a further unsprung road wheel next to the idler, and four return rollers. The tank was first displayed as a propaganda vehicle at the Berlin International Automobile Exposition in 1939.

(**Opposite**) The first of three images showing Walther von Brauchitsch, *Oberbefehlshaber* (Commander-in-Chief) of the German Army from February 1938 to December 1941, standing atop the first Nb.Fz., delivers a speech to armaments workers in May 1938. Note the more rounded Rheinmetall turret in contrast to the angular Krupp design. The main armament was a 7.5cm *Kampfwagenkanone* (KwK) L/24 main gun with a coaxial 3.7cm KwK 36 L/45 cannon in the main turret – side by side in the Krupp turret, one above the other in the Rheinmetall turret – plus two machine-gun turrets. The six-man crew consisted of commander, gunner, loader, driver and two machine gunners (in separate turrets). (Galland Books)

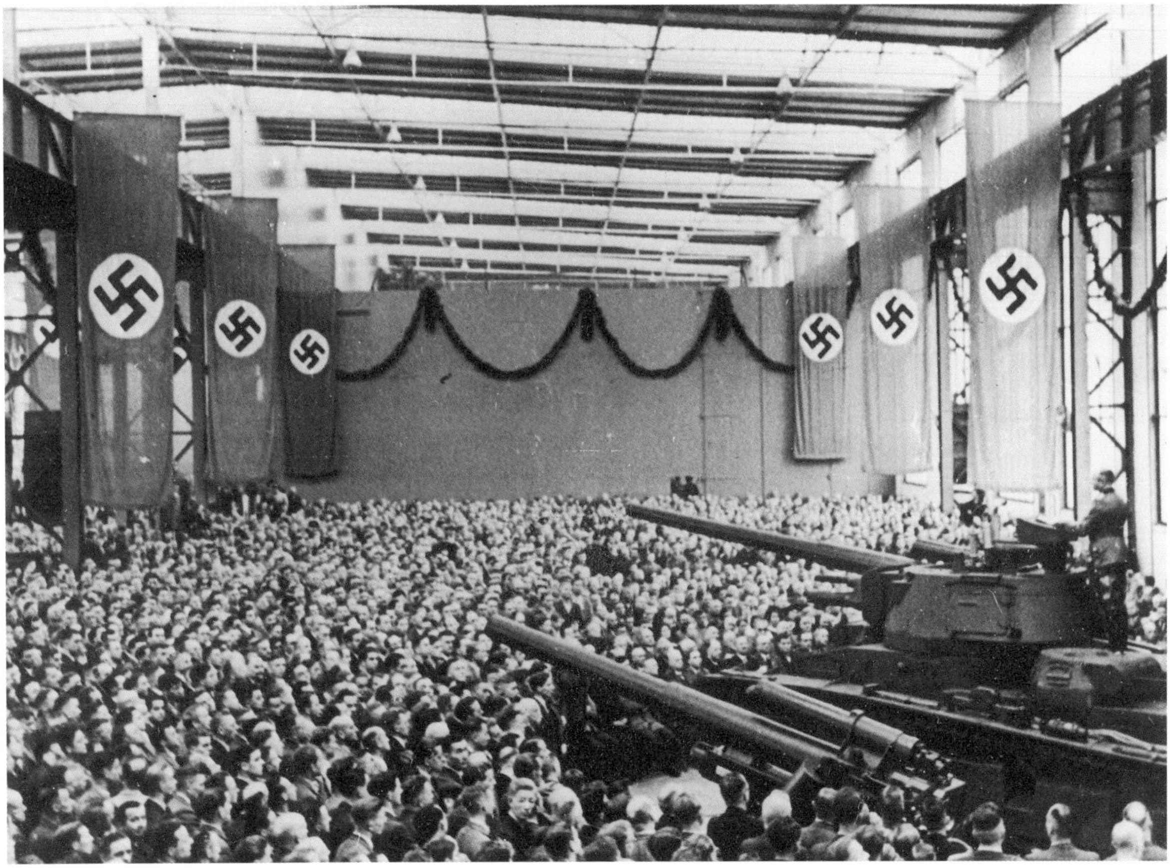

Three Nb.Fz. were assigned to Panzer-Abteilung z.b.V. 40 (*zur besonderen Verwendung*, 'for special duties') to support Operation *Weserübung*, the invasion of Norway in 1940, after transports carrying Pz.Kpfw. III and IV were lost at sea. The Nb.Fz. was the heaviest Wehrmacht German tank to see action in Norway. *Oberstleutnant* Ernst Volckheim, a former A7V *Sturmpanzer* commander and commanding officer of Panzer-Abteilung z.b.V. 40, favourably documented how the tank 'could be used in the mountains with exceptional success'. The multiple turrets, he recorded, 'were particularly useful in pinning down enemy infantry hidden behind roadside barricades'. One Nb.Fz. was knocked out by anti-tank fire, and a section of its running gear is today preserved in a small museum at Kvam.

Frames from newsreel footage of the Nb.Fz. in Norway. After parading through Oslo, the tank was attached to the 196th Infantry Division. One Nb.Fz. was immobilized by British anti-tank gunners at Kvam on 25 April.

Nb.Fz. and *Gebirgsjäger* (mountain troops). Note the Krupp main turret and and secondary turret — a modified Pz.Kpfw. I turret with a single 7.92 mm MG 34 air-cooled machine gun. The 23-ton tank measured 6.65m in length, 2.90m in width and 2.90m in height. Protected by up to 20mm of armour, it had a maximum speed of 30km/hr and a range on road of 120km.

Chapter Two

Pz.Kpfw. I (MG) (Sd.Kfz. 101) Ausf. A

The earlier testing at Kazan and retooling of German industry paved the way for large-scale tank production in Adolf Hitler's rearmament programme. Development of a new German light tank, a *Kleintraktor* (small tractor), was tabled in early 1930. Problems with the rear-driven *Leichttraktor* throwing its tracks, plus the influence of British Vickers-Carden-Loyd tractor suspension, led to the new tank having front drive with a rear-mounted engine.

The first *Kleintraktor* chassis was demonstrated in July 1932. Later trials demonstrated superiority over the British vehicle, prompting General Oswald Lutz to request an order for five *Kleintraktoren* with a machine gun-armed turret. Subsequent design improvements included wider tracks, a more powerful 60hp engine, a third return roller and shock absorbers for the front wheels. The first chassis was delivered in July 1933, the next four in August. Further improvements followed, with Krupp awarded a contract for an initial batch of 135 *Kleintraktoren* under the guise of *Landwirtschaftlicher Schlepper* (La.S) or 'agricultural tractor'. Another fifteen vehicles were to be produced by Rheinmetall, Henschel, MAN, Krupp-Grusonwerk and Daimler-Benz.

British Vickers-Carden-Loyd tractor (serial number VAE 393), purchased by *Waffen Prüfen 6* (Wa.Prüf.6) in November 1931. Subsequent testing assisted in the development of the *Kleintraktor*. Wa.Prüf.6 operated under the *Heereswaffenamt* (Army Ordnance Department) as a design office for tanks and motorized vehicles.

Prototype *Kleintraktor*, codenamed *Landwirtschaftlicher Schlepper* (La.S) or 'agricultural tractor'. The suspension (with two return rollers and five road wheels) was an original design and was not based on the Vickers-Carden-Loyd tractor. It was first demonstrated to Wa.Prüf.6 on 29 July 1932.

Plans for Krupp-Grusonwerk to upgrade the first series La.S with a superstructure and turret were cancelled after the armour plate of the Krupp vehicles failed testing. Overhauled, these chassis (with the additional of a rail) became training vehicles. (Galland Books)

Work on a turret and superstructure (from Daimler-Benz) continued, using a mock-up (pictured here) produced using a first series La.S chassis. Sheet-metal turrets and superstructures were subsequently added to twenty-first series chassis.

A first series La.S with armament-less Krupp turret. The suspension comprised a front sprocket, four road wheels, three return rollers and an idler (flush with the ground). The first road wheel has a coil spring and shock absorber; the second and third road wheels are paired in a suspension cradle with leaf springs pivoting on the second axle; the fourth road wheel similarly pivots on the third axel and is connected to the idler via a suspension cradle. An outer U-rail connects the second and third axels.

Discussions began in July 1933 for the second La.S production series. With development of the *Landwirtschaftlicher Schlepper* 100 (later to become the *Panzerkampfwagen* II) behind schedule, priority was given to La.S manufacture. With secrecy no longer required, the tank was renamed MG *Panzerwagen*, then MG *Kampfwagen*, before officially becoming the *Panzerkampfwagen* I (Pz.Kpfw. I). The first serial model was designated *Ausführung* (Ausf.) A. (Galland Books)

Pz.Kpfw. I Ausf. A on a Rheinmetall assembly line in October 1935. Designated as *Sonderkraftfahrzeug* 101 (special motor vehicle), or Sd.Kfz. 101, the *Panzerkampfwagen* I (Pz.Kpfw. I) was the first German tank to enter mass production since 1918. Note the two-part *Einsteigluke* (driver's hatch) on the left side of the vehicle. The tank was an important step in the development and expansion of German industry. Monthly production of sixty vehicles in 1935 increased to seventy the following year. A total of, 1,075 Pz.Kpfw. I Ausf. A were manufactured across four series including fifteen exported to China.

(**Above**) The Pz.Kpfw. I Ausf. A was jointly manufactured by Maschinenfabrik Augsburg-Nürnberg AG (MAN), Henschel & Sohn GmbH (Henschel), Daimler-Benz and Fried. Krupp Grusonwerk AG (Krupp-Gruson). Measuring 4.02m long and 2.06m wide, the tank weighed 5.4 tons. (NARA)

(**Opposite**) The height of the tank, 1.71m, is apparent beside the two-man crew: driver and commander/gunner. Note the original long-nozzle Bosch horn; this was replaced in 1935 by a smaller cylindrical version. Armour plate up to 13mm in thickness provided protection against steel-cored S.m.K. armour-piercing ammunition. (NARA).

(**Below**) A pause in training. The two *Maschinengewehr* (MG) 13 machine guns in the manually traversed turret were supplied with 2,250 rounds of 7.92mm ammunition in twenty-five box magazines. (NARA)

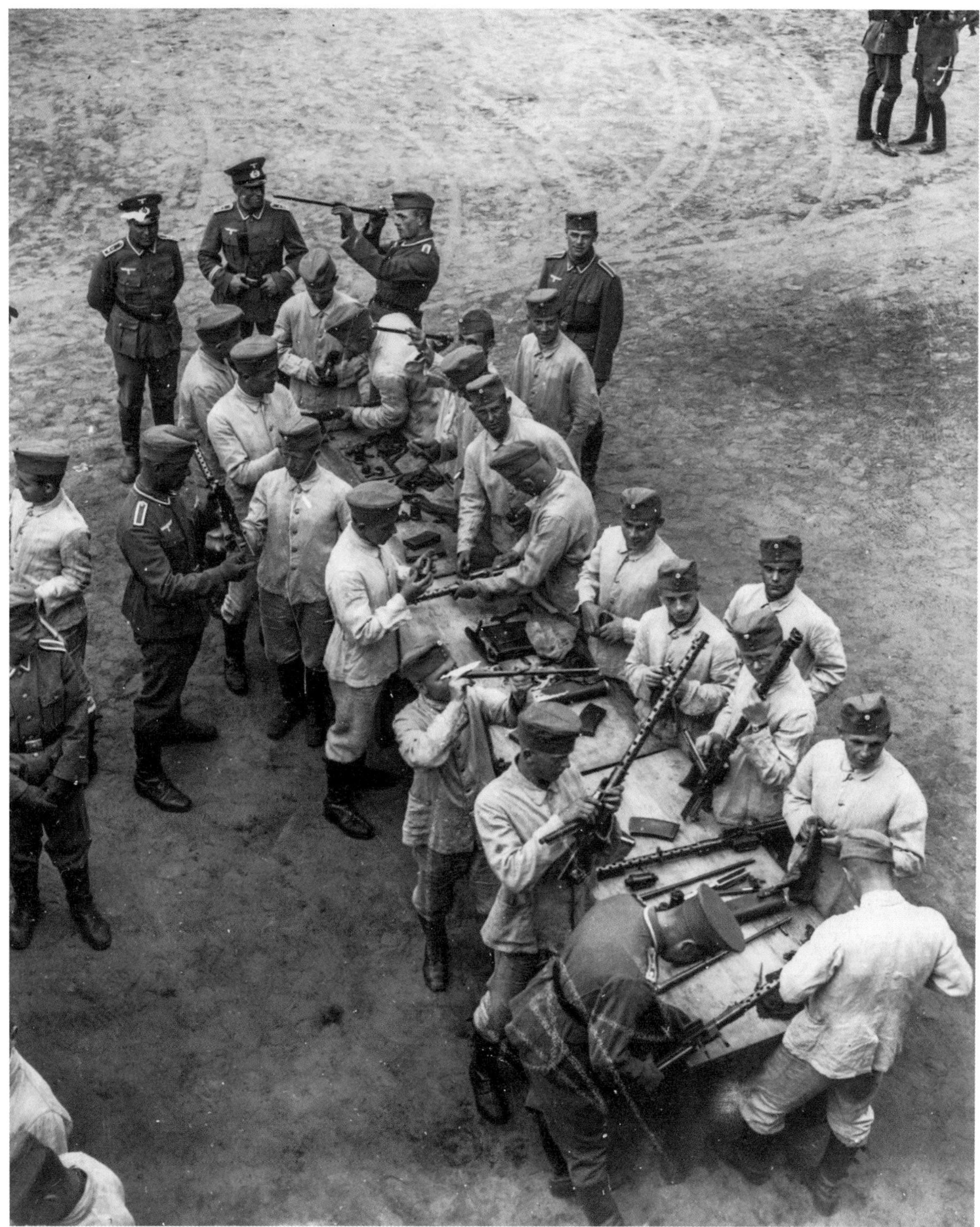

Two images showing supervised servicing of the MG 13. The MG 13, a short recoil, air-cooled light machine gun, was adopted by the German Army in 1932. To disguise the clandestine development of the weapon from Allied weapons inspectors enforcing the Versailles Treaty, its designation referred to the older 1913 pattern weapon used during the First World War. Note the wearing of *Reichswehr*-issue *Krätzchen* ('pork pie') caps, which preceded the later Wehrmacht side cap. (NARA)

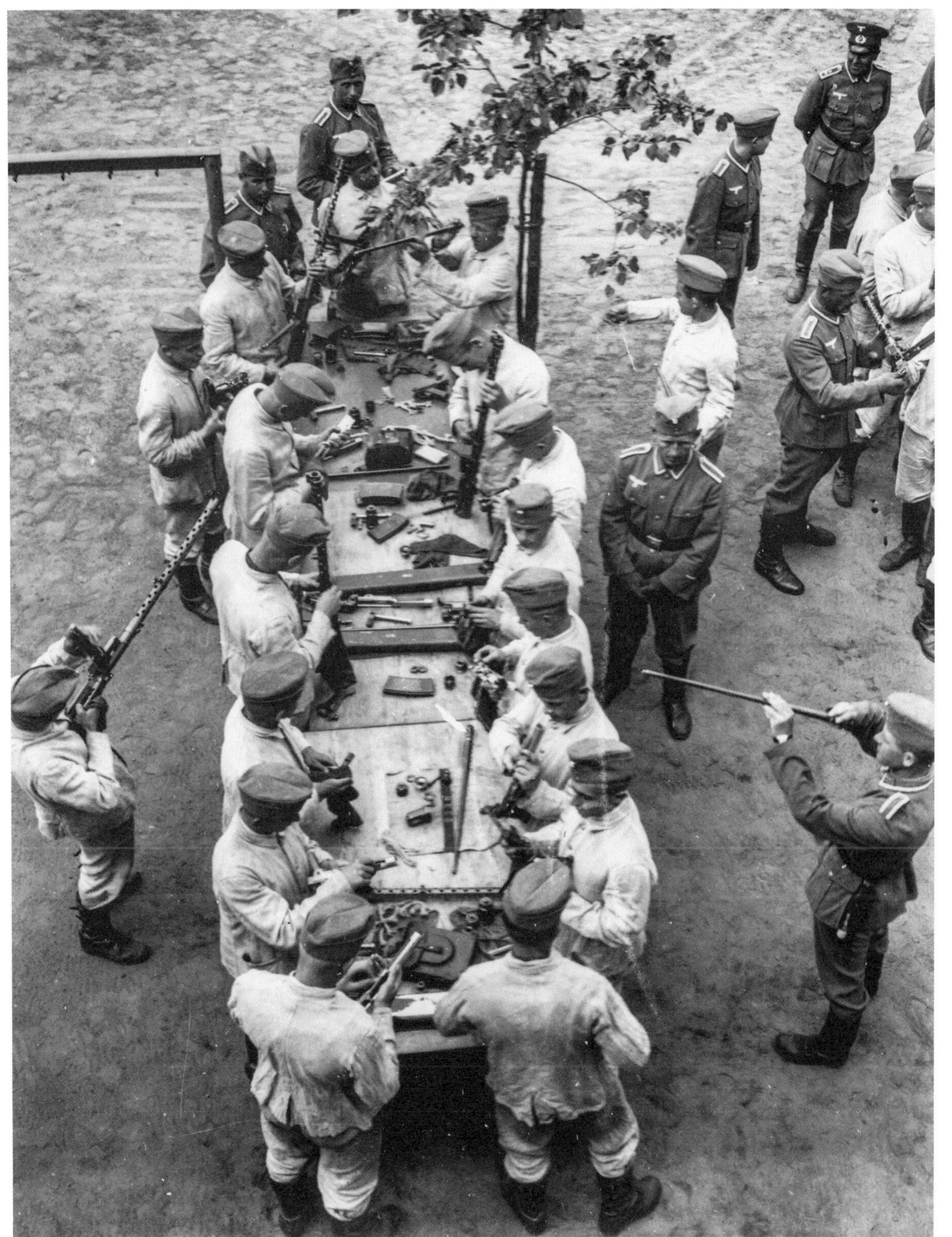

Finnish soldier atop a Pz.Kpfw. I Ausf. A turret. The staggered arrangement of the MG 13 machine guns facilitated the changing of the 25-round drum magazines within the confined space of the turret. The left gun mount was attached to the gun mantle while the right gun mantle was adjustable. Note the turret lug, one of three, for lifting the turret or holding camouflage materials. (SA-Kuva)

MG 13 cleaning and maintenance during a break in manoeuvres. (Christian Ankerstjerne's collection)

Panzer-Regiment 1 Erfurt

The 1st Panzer Regiment, one of two Panzer regiments within the 1st Panzer Division, was formed on 15 October 1935. The regiment's barracks were located in Löbervorstadt, a suburb of Erfurt in the central German state of Thuringia.

Period postcard offering 'greetings from Panzer Regiment Linten: Our faces are dusty, but our spirits are happy.'

5th Panzer Regiment

Three images showing Pz.Kpfw. I Ausf. A from the 5th Panzer Regiment passing a saluting local crowd and marching band as they approach the Wünsdorf garrison after a road march along Reich Highway 96 from Zossen on 20 October 1935. (NARA)

Hauptmann Wolfgang Thomale, commander of the 1st Company, 1st Battalion of the 5th Panzer Regiment, flies a *Totenkopf* (death's head) pennant. (NARA)

Elements of the 5th Panzer Regiment enter Garrison III, later christened Cambrai Barracks, on 22 February 1938. (NARA)

Conscription was reintroduced in Nazi Germany on 16 March 1935. The following series of images depicts the swearing-in of new recruits in November that year. The majority of men who became *Panzertruppen* were selected on merit, second only to the Luftwaffe in general aptitude. (NARA)

(**Above**) Amidst rearmament and army expansion, a monument to the A7V was erected at the 5th Panzer Regiment's Cambrai Barracks at Wünsdorf. A brass plaque beneath read: 'In the Spirit of Comrades from the World War: Attack – Fight – Win.'

(**Opposite, above**) A replica A7V, named *Hummel*, was also put on display at the Berlin *Zeughaus* (armoury) during the 1930s. An iron cross was later painted on the front of the wooden mock-up.

(**Below**) Hitler, who had written of the importance of motorization in *Mein Kampf*, was excited by the potential of the tank. After viewing a 1936 training exercise, he remarked: 'That's what I need, that's what I have to have!' (NARA)

(**Above**) Note the rectangular *Sehklappen* (vision ports), each 4mm wide with a 12mm laminated glass block providing eye protection. (NARA)

(**Opposite**) The first *Panzereinheit* (Panzer unit) at Wünsdorf in 1935. Each *Panzer-Zug* (platoon) was initially assigned seven Pz.Kpfw. I, a number later reduced to five. It was anticipated that the 1st and 2nd Panzer Divisions would be combat ready in April 1936, the 3rd Panzer Division eighteen months later. Larger tanks equipped with 3.7cm and 7.5cm guns (which became the Pz.Kpfw. III and IV) would follow. (NARA)

Blackboard gunnery training showing the arc of a fired projectile and reticle of the *Turmzielfernrohr 2* (TZF2, Telescopic Sight 2) gunsight, with a magnification of × 2.5 and field of view of 28° over 800m. The reticle could be seen in the centre of the sight, with a large *Hauptstachel* (triangle shape) and three smaller triangles on each side to estimate the height and width of the target. (NARA)

Note the reticle on the top left of the blackboard. Once the range was estimated, the gunner would adjust the range of the guns and aim using the top of the central *Hauptstachel*.

Friedrich Sander, 11th Panzer Regiment, detailed his initial training on the Pz.Kpfw. I:

For the recruits, the average day began at 5.30am with early morning sports. This was followed by time for bodily hygiene (with cold water) and breakfast. At 7.00am the theory began, given by the officers of the regiment: the soldier's oath, duties of the soldier, command and obedience, cartography, structure of the army, counterespionage, discipline and disciplinary rules and similar topics were covered. Then followed NCO lessons on a variety of topics such as: behaviour towards superiors, saluting, rank insignia, weaponry, dress code, shooting practice, equipment, first technical instructions and the like. A short break was followed by drill, without and with carbine 98, weapons drill, shooting practice, general infantry service, marching and cross-country exercises. This duty on the parade ground and in the camp was only interrupted by a ninety-minute lunch break and usually ended between 4.00 and 5.00pm. This was followed in the barracks by weapon cleaning, cleaning and mending sessions and repeated room, uniform and locker inspections which ended between 6.00 and 8.00pm. After the first five weeks, there was training on the tank as a driver, gunner or radio operator ... Those destined to become NCOs or officer candidates received additional marksmanship lessons, tactical instruction and general schooling in a variety of subjects. Tactical instruction was conducted by the company commanders or one of the company officers and consisted of, among other things, cartography, preparation of reports and sketches, firing from the tank on the training ground, firing on the move, fighting methods of the armoured combat vehicles, combined arms tactics, sandbox exercises, command techniques and additional terrain training. In addition, the NCO candidates received general instruction in politics, history, party affairs, racial politics and other subjects and every week there was a fifteen-minute special lecture to train and consolidate personal conduct in front of comrades and superiors.

Driver training. Note the twin steering levers and gear shift lever.

Nationalsozialistisches Kraftfahrkorps (NSKK, National Socialist Motor Corps) driving instruction. This paramilitary organization, formed on 1 May 1931, aimed to educate its members in *Motorische Ertüchtigung* (motor training).

The first of two images showing changing a track on a Pz.Kpfw. I Ausf. A under a watchful eye. The unlubricated tracks were joined with a pin and secured by a wire bent in an s-shape. Early pattern tracks featured triangular cavities in the guide horns to help reduce the accumulation of earth, which could throw a track. Later pattern tracks featured solid guide horns. (Christian Ankerstjerne's collection)

The rear deck of a Pz.Kpfw. I Ausf. A with the twin exhaust pipes (superseded by a single, rear-mounted exhaust muffler on the Ausf. B and redesigned armoured access hatches) leading from the 57hp, four-cylinder, air-cooled Krupp M305 engine. Note the commander's TZF2 gunsight inside the turret. (Galland Books)

Pz.Kpfw. I Ausf. A participating in the 8th *Reichsparteitag* parade. These early model tanks lack a vision slit in the side vision ports. Note the commander communicating with the driver via a *Sprechschlauch* (speaking tube) and the 3-ton jack mounted on the track guard.

A company commander's Pz.Kpfw. I Ausf. A during a 1935 *Reichsparteitag* parade showing the reverse of the *Totenkopf* (Death's head) pennant. Writing on the utility of light tanks, *General* Heinz Guderian concluded in his 1937 book *Die Panzertruppen*: 'Their low target height, high speed, and manoeuvrability make them an especially dangerous enemy of anti-tank guns … Don't underrate these opponents when they are employed in large numbers.'

Panzertruppen were an elite force, and received a high level of training and a distinctive black uniform. The two-part M1934 *Schutzmütze* ('Panzer Beret') comprised an inner crash helmet concealed by a removable black woollen cover. (NARA)

Panzertruppen postcards depicting a later version of the *Schutzmütze* featuring an embroidered swastika and eagle badge above a cockade of the national colours within an oak leaf wreath.

The double-breasted black wool field jacket, introduced in 1934, featured aluminium *Totenkopf* insignia, pink piping around the collar tabs and shoulder boards, and a breast eagle.

The Pz.Kpfw.I was Germany's most numerous prewar tank, a crucial weapon in building the *Panzerwaffe* and shaping its doctrine.

Thirteen propaganda postcards portraying the Pz.Kpfw. I Ausf. A as a capable all-terrain vehicle. In fact, the underpowered tank had an average road speed of 20km/hr and 10–12km/hr across country.

'The slope is steep, but it is no obstacle for the tank.' The following are propaganda images of the Pz.Kpfw.I Ausf.A taken at a training ground. While the tank strikes a menacing pose, its exposed underbelly was protected by a mere 6mm of armour. The recessed front hull was designed to aid manoeuvrability in uneven terrain.

Prewar postcard pictures of Pz.Kpfw. I Ausf. A traversing a river. It appears that the nearest Pz.Kpfw. I Ausf. A has stalled, with the driver outside the tank and steam rising from the engine compartment. (Galland Books)

Adapted from a Kriegsmarine song by Oberleutnant Kurt Wiehle in June 1933, the *Panzerlied* (lit. 'tank song') became the *Panzerwaffe* marching song: 'Whether it's freezing or snowing, whether the sun is shining on us; the day is blazing hot or the night is dark; our faces are dusty, but our spirits are happy; our tank roars along in the stormy wind.'

Frequently dismissed as a training tank, indeed derided postwar by Guderian as just that, the Pz.Kpfw. I was designed to be an offensive weapon. German planners estimated that an armoured battalion of 100 tanks could break through a French infantry division (armed with seventy-two anti-tank guns) with 50 per cent losses – a casualty rate substantially lower than on a First World War battlefield. (NARA)

Turning heads was this armament-less *Kleiner Panzerbefehlswagen*, a command variant of the Pz.Kpfw. I with rectangular superstructure in place of the MG turret to accommodate radios and a radio operator (see page 168).

A Pz.Kpfw. I Ausf. A with a single machine gun is examined with a second tank concealed beside it.

German tanks in the period 1935 to 1937 were painted in a three-tone *Buntfarbenanstrich* (colourful paint) camouflage pattern comprising *Grün-matt* (green), *Erdgelb-matt* (earth yellow) and *Brau-matt* (brown). (NARA)

Two views clearly showing the distinctive *Buntfarbenanstrich* camouflage. The twin machine guns of this tank are fitted with a *Rückstoßverstärker* (recoil booster) to allow automatic fire during a training exercise since blank ammunition provided insufficient recoil to allow automatic fire. Note the early long nozzle horn on the front right of the tank. The Daimler Benz-designed turret could traverse through 360° using a hand wheel with one rotation swinging the turret by 6°. The gunner could also disengage the gearing and manually swing the turret. (NARA)

Pz.Kpfw. I Ausf. A on parade in front of Nürnberg Hauptbahnhof (train station) en route to the Zeppelinfield, site of the Nazi Party Rally Grounds.

Playing card symbols denoted formations from the 1st Panzer Division during their *Lehr-und-Versuchsübung* (teaching and experimental exercise) at Munster in July or August 1935. (NARA)

Pz.Kpfw. I Ausf. A (La.S second series) with exposed engine deck before the addition of armoured covers. Note the vision port on the right rear superstructure, which was discontinued after the first 300 examples, and the absence of additional side armour.

The Pz.Kpfw. I Ausf. A had a fording depth of 60cm. These pre-war tanks, with a rhomboid-shaped tactical marking on their turret, did not require a blackout cover on the central headlight. (Christian Ankerstjerne's collection)

Pz.Kpfw.I Ausf.A from the 2nd Panzer Division. Underscoring the need for training exercises, Guderian wrote in the 1930s that 'fire control and a high standard of tank gunnery training are the factors that will contribute most towards victory'. As a junior officer, Hans von Luck recalled how 'Guderian visited every single company, observed its training, and afterwards discussed his ideas with its officers and NCOs. We realized that in addition to training, material and modern technique, the spirit that inspired a unit also played an important part.'

After the aforementioned exercises in August 1935, *General der Panzertruppe* Oswald Lutz concluded: 'Only the employment of Kampfwagen in depth promises success. About three kilometres wide is viewed as the smallest front for employment of an entire Panzer Division … The Machine-Gun-Kampfwagen are a very useful and effective weapon.

Pz.Kpfw. I Ausf. A during peacetime exercises. Note the bottom left tank, bogged and flying a breakdown flag.

Pz.Kpfw. I Ausf. A and a *leichte* (*Funk*) *Panzerwagen* from the 6th Panzer Regiment (note the symbol on the superstructure behind the turret) on parade through the town of Neuruppin in Brandenburg on 14 June 1936.

A mechanized parade marks Hitler's 47th birthday on 20 April 1936. Here Germany's Führer salutes passing Pz.Kpfw. I Ausf. A. outside the *Technische Universität Berlin* (Technical University in Berlin). In a radio address on the eve of his birthday, Dr Joseph Goebbels proclaimed: 'Tomorrow morning, this whole people wishes to proclaim its love and honour to the Führer, but also its thankfulness for his impact on humanity and on history.' Readers of a London newspaper were informed that 'such a parade of military power has not been seen, even in the most spectacular days of the Kaiser'. (Galland Books)

Nürnberg

The Nürnberg rallies, held in 1923, 1927 and 1929, and then annually from 1933 to 1938, were a carefully choreographed showcase of Germany's nascent military forces, an occasion where Adolf Hitler could foster *Volksgemeinschaft* (a unified people's community) as he propelled his country towards a new war. Massed tank parades demonstrated the Wehrmacht's growing might. (NARA)

Pz.Kpfw. I Ausf. A from the 1st Panzer Division on display at the eighth Reich Party Congress, known as the *Reichsparteitag der Ehre* ('Rally of Honour'). Writing of the 'Zeppelin Field, completed for the 1936 Parity Congress of Honour', Dr Walther Schmitt, in the official NSDAP book, described a 'shimmering row of columns 350 metres long leads from the large platform from which the Führer speaks. The proud bright marble joins with the brilliant red of the swastika flags to form an indescribably festive harmony.' A London newspaper reporting on the congress noted a 'demonstration of the latest methods of trench warfare, revealing that Germany has trebled her heavy [*sic*] tanks and other equipment within a year. The display culminated in the Minister for Defence (General von Blomberg) presenting the army with new colours and saying: "Your highest duty is to follow them to death or victory."' (NARA)

Regarding the tank parade, the official congress programme, *Der Parteitag der Ehre vom 8. bis 14. September 1936. Offizieller Bericht über den Verlauf des Reichsparteitages mit sämtlichen Kongreßreden*, trumpeted: 'The 1. Abteilung of a tank division begins manoeuvres. Entering from the central gate opposite the main platform, they spread out before the platform and conduct various manoeuvres. They form two columns, and weave back and forth, which requires exact timing. Forward and diagonal manoeuvres follow. At the command that battle is near, the heads of the gunners disappear from the turrets. One hundred tanks fire a thundering salvo. The tanks roll past the main platform and leave the field.' (NARA)

(**Above**) A simulated attack inside the rally ground. According to the programme: 'The last part of the exhibition is a large mock battle. A strong defensive position with machine guns and light artillery is established to the west. It is defended with barbed wire and mine fields. The attackers – the III. Infantry Regiment 11 – enters from the east. The attack begins, supported by heavy machine guns and artillery. The enemy answers with heavy fire against the machine guns. The attackers work forward through heavy fire. But the enemy position is too strong. Tank units come to the support of the infantry, opening heavy fire on the enemy position. The hellish noise reaches its high point. Moving forward through the infantry, the tanks advance despite heavy fire. They reach the minefields. Mines explode, disabling some tanks. The second wave succeeds in reaching the enemy position. They roll over the barbed wire. The enemy retreats, and the riflemen follow behind the tanks to pursue the enemy. After heavy combat, the enemy position is taken.' (NARA)

(**Opposite**) German military might on display to the world. After a final parade, Hitler addressed the assembled troops, recounting the bitterness of defeat in 1918 and the Versailles Treaty, and restoring a new confidence in the nascent Wehrmacht: 'In a few weeks it will be exactly eighteen years since the great army, the proudest symbol of our people, struck by treacherous assault, had to lower its arms and flags …' The massive grandstand with a width of some 400 metres was designed by Adolf Hitler's favourite architect Albert Speer, who drew inspiration from the Pergamon Altar built (in modern-day Turkey) by the Ancient Greek King Eumenes II.

(**Above**) Mock combat. British Major General Arthur Cecil Temperley, writing as a military correspondent, praised 'the excellence of the arrangements and completeness of the [German] equipment … a tribute to the progressive thinking of the German General Staff.'

(**Opposite**) Three propaganda images displaying 'the latest weapon of the Wehrmacht. The tank has passed through a forest and now the tank gunner is scanning for the enemy.' Tank commanders were taught three defensive fighting positions: open, hull-down and concealed. A concealed or camouflaged position, such as the edge of a wood, could thwart enemy observation but without protection against enemy fire. (NARA)

In the absence of radio in early Pz.Kpfw. I, flags were used to relay commands. Red, yellow and blue flags were used singularly and in combination to relay commands including orders to 'close hatches', 'attack' an enemy tank or gun, or 'stand down'. A round port incorporated into the commander's hatch allowed signalling when the hatch was closed. (NARA)

The Pz.Kpfw. I was later fitted with a FuG 2, a radio receiver that enabled the commander to receive orders and information from command tanks.

Coming to grief during manoeuvres in Mecklenburg (Northern Germany) staged 18–24 September 1937. The Pz.Kpfw. I Ausf. A was part of the Blue party (containing elements of the 3rd Panzer Division; 1st Panzer Brigade; 3rd, 12th and 23rd Infantry Divisions) with the objective of breaking through the Red party (containing elements of the 30th Infantry Division). Note the *kleiner Panzerbefehlswagen* in the background.

This immobile Pz.Kpfw. Ausf. A., '8', displays an *Ausfallflagge* (breakdown flag) on the turret. The flag, 40cm by 30cm, consisted of a black cross on a yellow background. Note the older style First World War *Stahlhelm* (steel helmet) worn by the surrounding infantry. (Christian Ankerstjerne's collection)

'Light tanks from four to seven tons', such as the Pz.Kpfw. I, Guderian surmised in *Die Panzertruppen*, 'armed only with machine guns, are quite often used for the tasks of reconnaissance, security, and running orders. In addition, particularly while supported by heavier [vehicles], these light combat vehicles are quite suitable for engaging infantry and other troop targets.'

Barracks shot. An NCO in linen *Drillichanzug* (fatigue uniform) beside his Pz.Kpfw. I Ausf. A. The colour of the *Leinen-Drillich* (herringbone) linen tended to quickly fade. (Galland Books)

Hatches and visors are open in this Pz.Kpfw. I Ausf. A. Note the driver's two-part hatch; the upper section opened first and closed last. (Galland Books)

The driver's view from his interior position on the left side of the tank (above) and the commander's view (bolow). Although it is not clear from these images, German tank interiors were painted in *Elfenbein* (ivory) and *Graugrün* (grey-green), the latter being more resistant to dirt and oil. (NARA)

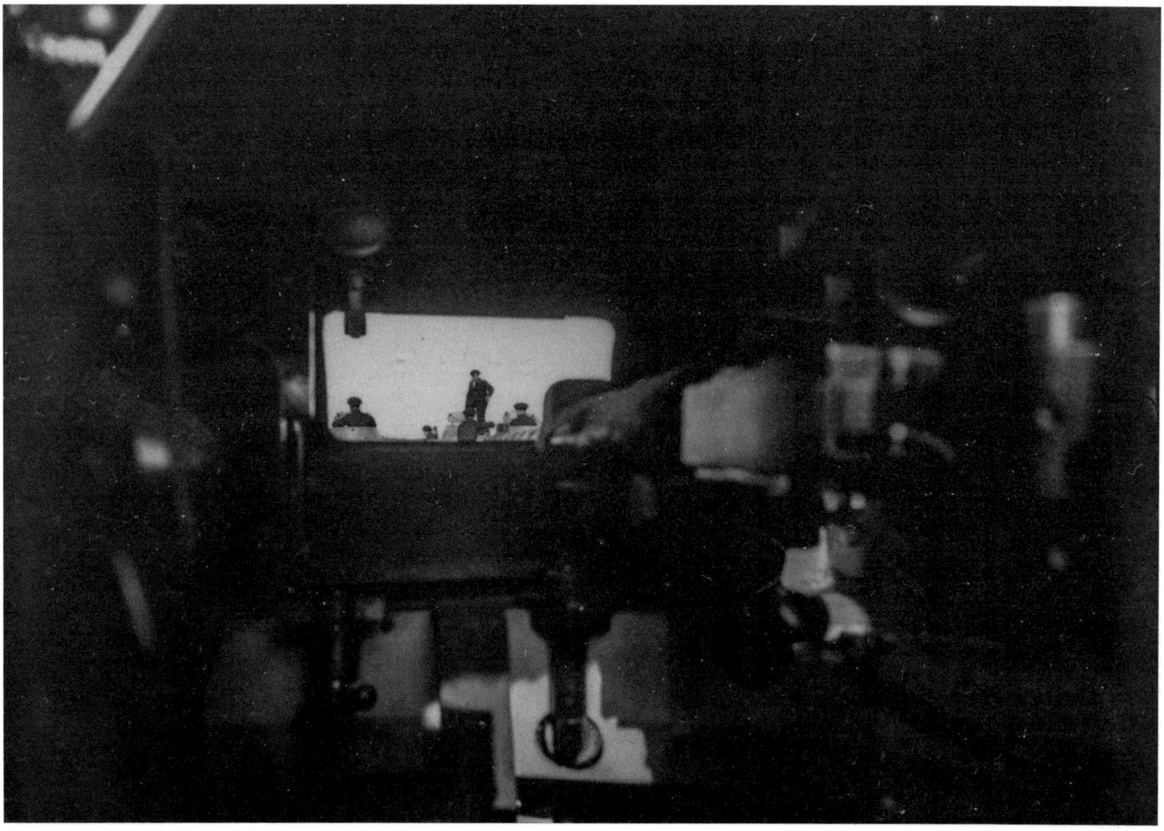

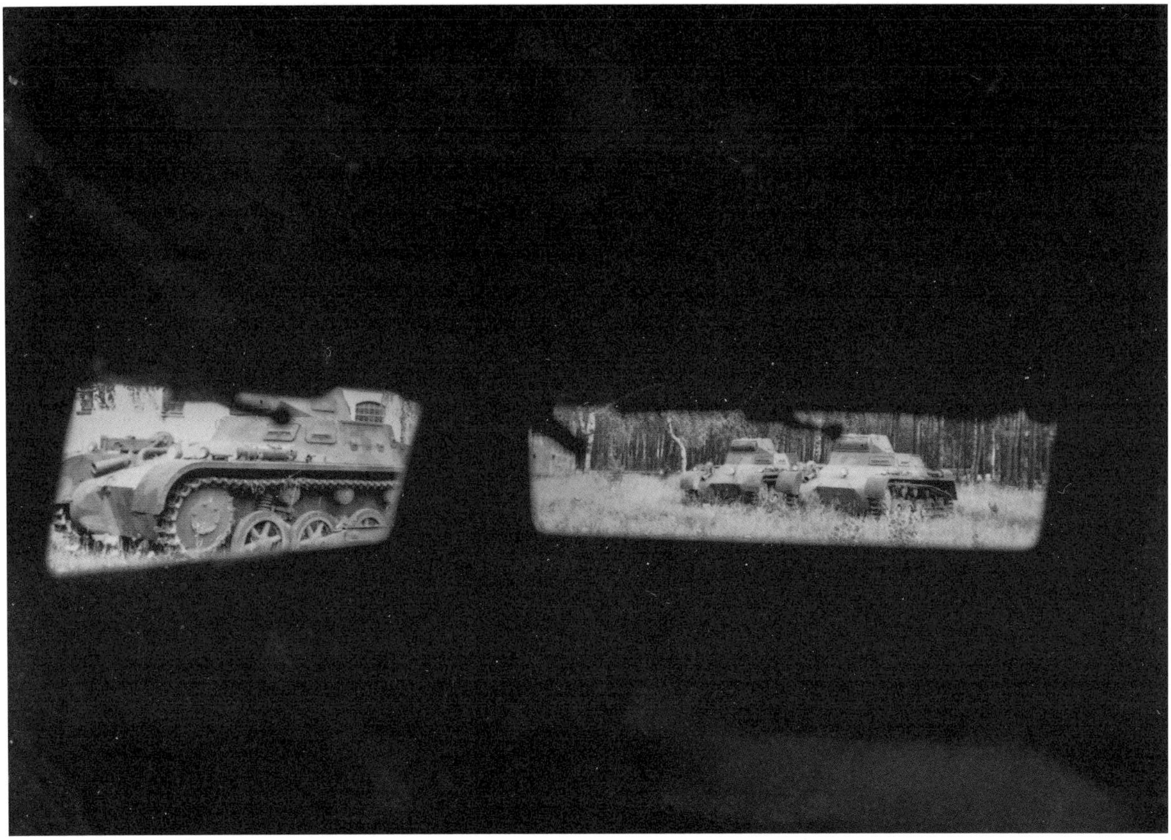

The view with the forward vision ports open. These images were taken in late July or August 1935 at *Truppenübungsplatz Munster* (Munster training area) in northern Germany. (NARA)

Peacetime training at Wünsdorf. Drivers were taught to negotiate ditches front on, rather than approaching at an angle.

Pz.Kpfw. I Ausf. A of the 2nd *Lehrtrupp* of *Kraftfahrlehrkommando* I Zossen, identified by the triangle with two horizontal lines on the rear of the turret, during a demonstration.

Hitler and his entourage arrive for the *Reichserntedankfest* (the Reich Harvest Thanksgiving Festival) – an annual Nazi celebration of German farmers that also showcased the country's expanding military. It was held from 1933 to 1937. Around a million people attended the 1936 event, up to 1.2 million in the following year. The event was designed by architect Albert Speer, who had earlier written that 'The material with which to create a mass event, first and foremost, is always the crowd of people themselves.' Running through the centre of the crowd was the *Mittelweg* (central pathway), measuring 800m long and connecting the large upper podium for *Ehrengäste* (guests of honour) with the smaller podium closer to the camera, Hitler's *Rednerkanzel* (speaker's pulpit). (NARA (top))

Armour and infantry pass through a mock village, soon to be contested.

(**Above**) Newsreel footage of an MG *Panzerwagen* at the *Reichserntedankfest*.

(**Opposite**) In a scene from the *Schauübung* (military show), German troops fight for control of Bückedorf, a mock village erected by pioneers.

Further four images from the *Reichserntedankfest*, with Bückeberg, a hill near Hamelin (Lower Saxony) visible in the background. The sloping hillside provided a natural amphitheatre that could accommodate hundreds of thousands of spectators. Speer surrounded the arena with a horseshoe-shaped ring of flags to give those assembled 'the feeling of belonging' (Speer 1933) or *Volksgemeinschaft* (national community). The military *Schauübungen* (demonstration shows) grew more elaborate in each year. Hamelin liked to call itself the *Nürnberg des Nordens* (Nuremberg of the North). (NARA)

The 3rd series tank (above) on the right rear is fitted with a brake light but no reinforcing side armour. (NARA)

Civilians and soldiers mingle among the tanks, in a spectacle designed to impress spectators, *Volksgenossen* (comrades), as well as the Führer. In his 1935 speech, Hitler boasted: 'Destiny has made it possible for us this year, not only economically to yield a rich harvest, but it blessed us with more: the German military armed forces have risen again. The German fleet will arise again. German towns and beautiful villages, they are protected, the power of the nation is watching over them, the force in the air.' (NARA)

Tanks outside their garage, Bückeberg hill in the distance. (Christian Ankerstjerne's collection)

The barrel cover on the closest Pz.Kpfw. I Ausf. A denotes *Fahrgestell* 10217, which identifies the tank as a fourth series vehicle manufactured by Henschel. (NARA)

Pz.Kpfw. I Ausf. A on display. 'The attack by tanks', Heinz Guderian wrote in 1936, 'must be conducted with maximum acceleration in order to exploit the advantage of surprise, to penetrate deep into enemy lines, to prevent reserves from intervening, and to extend the tactical success into a strategical victory. Speed, therefore, is what is to be exacted above anything else from the armoured weapon.' A total of 1,160 of these tanks were listed as available on 1 October 1936. (NARA)

Pz.Kpfw. I Ausf. A supported and 'attacked' by Luftwaffe aircraft during military exercises in 1937.

The 2nd Battalion, 5th Panzer Regiment organized a *Panzersportfest* (armoured sports celebration) near Wünsdorf on 28 June 1936. A host of events, including massed synchronized gymnastics, acrobatics on motorcycles, chariot racing and tanks smashing through obstacles, was held before some 30,000 spectators. Pictured is a Pz.Kpfw. I Ausf. A driving through a specially erected brick wall. A wooden barricade similarly presented no obstacle.

'The secrets of a combat vehicle' reads the postcard caption.

'The secrets of a combat vehicle' … recruits are similarly interested.

Military manoeuvres, September 1936

September 1936. Adolf Hitler and his entourage arrive to observe the field manoeuvres of the Fifth Army Corps, held between Frankfurt and Fulda on Reichsstrasse 276 (today Bundesstrasse 276) from Wächtersbach to Schotten. The demonstration took place near the village of Wächtersbach, 13km south of Birstein. Six months earlier Hitler had violated the Treaty of Versailles and the Locarno Pact by sending German military forces into the demilitarized Rhineland in western Germany. (NARA)

Briefing Hitler on the 'opposing forces'.

Supporting infantry use a *Maschinengewehr* 08/15 machine gun, a cumbersome 18kg, water-cooled, fully automatic weapon introduced onto the battlefield in April 1917. The drum magazine contained a 100-round cloth ammunition belt.

Hitler in discussion regarding the day's manoeuvres. *Generaloberst* Werner von Fritsch and *General-feldmarschall* Werner von Blomberg (on Hitler's right) were 'implicated' in 1938 scandals, leading to the formation of the *Oberkommando der Wehrmacht*, which was directly subordinate to the Führer.

Instructions are given using a standard Wehrmacht signalling disc.

Infantry cooperation. Pz.Kpfw. I Ausf. A passing a 7.5cm *Feldkanone 16 neuer Art* (7.5cm FK 16 nA), a First World War era 7.7cm gun re-barrelled during the 1930s. (NARA)

Inquisitive onlookers inspect Germany's first mass-produced tank. (NARA)

Engine hatches were kept open to ventilate the tank's four-cylinder air-cooled engine during and after a long march. (NARA)

Camouflaged Pz.Kpfw. I Ausf. A. Reporting on the manoeuvres, a London masthead reported: 'For the first time since the Great War British journalists attended the German military and aerial manoeuvres … The operations were on the largest scale held in Germany since 1914, and amazing progress was revealed. Every part of the new German Army was brought into action, but chief interest centred in the tank and other mechanised units in co-operation with the air-arm and anti-aircraft batteries. The publicity and stage management suggest that the manoeuvres may be regarded as a demonstration to the world of Germany's prowess, and also as an encouraging spectacle for the German people, rather than as the usual problem exercises.' (NARA)

Preparing for war. The massed use of tanks, General Heinz Guderian believed, was akin to 'hitting someone with your fist and not your fingers.' (NARA)

A period cigarette card from Garbáty Cigarettenfabrik. This Berlin-based company was denounced as a 'Jewish firm', forcing owner Moritz Garbáty to sell it in 1938. The description on the reverse of the card reads: 'The tracked drive enables the *Panzerkampfwagen* to drive on almost any terrain, climb steep inclines, and break through even strong obstacles.'

Prewar postcards showing mock combat and firepower displays. In later combat, tanks would be spaced at least 100m apart with parade-like formations prohibited as Allied air power became more threatening. Tanks would only return to column formation in narrow areas or to bypass obstacles. A basic rule also taught commanders to avoid using a single road in order to spread enemy fire.

Dramatic propaganda image of a Pz.Kpfw. I Ausf. A. Note the damage to the rubber on the front road wheel. (Christian Ankerstjerne's collection)

Raised using a 3-ton jack, the rear idler of this Pz.Kpfw. I Ausf. A is examined for damage.

Additional armour, bolted beneath the Pz.Kpfw. I Ausf. A side entry door, featured on both sides of later production vehicles.

Pz.Kpfw. I Ausf. A crews honing their skills. Attending a 1936 lecture entitled 'The Bogey of the tank', given by a German engineer to the so-called 'Society of Popular Education', a British journalist described how: 'In 1917 and 1918 it [the tank] became the most destructive weapon on the Western Front.' Building on the 'shackles of Versailles', the speaker described 'the advantages and disadvantages of all chief tank types in use to-day … Lastly, amid cheers, the new Nazi tank units were flashed on the screen.' Finally, 'German genius', he boasted, 'has not allowed us to fall behind.'

Press photograph issued in America in June 1941 showing troops with fixed bayonets 'and a small tank' during the German invasion of the Soviet Union. The photograph was most likely taken in Germany during pre-war exercises.

An understudy Pz.Kpfw. II

A 1937 directive for the training of armoured troops stated that in the absence of Pz.Kpfw. II, a substitute Pz.Kpfw. I would carry a distinctive black-and-white checkerboard pattern around the top of the turret. Regulations at the time specified the roles different model tanks would play: the Pz.Kpfw. I would engage live targets; the Pz.Kpfw. II would engage both live targets and enemy armoured vehicles and artillery; the Pz.Kpfw. III would fight live targets with its three machine guns and enemy armour with its 3.7cm gun; and the Pz.Kpfw. IV would train its 7.5cm gun on important targets while supporting the advance of the lighter model tanks. Seen below is a Pz.Kpfw. I Ausf. A and *Umsetzfahrzeug* (bottom) in the guise of a Pz.Kpfw. II. (Christian Ankerstjerne's collection)

The checkerboard pattern squares on an 'acting' Pz.Kpfw. II measured 5.5cm by 5.5cm. A yellow and white turret band was applied to tanks masquerading as a Pz.Kpfw. III, while a solid yellow band represented a Pz.Kpfw. IV. (Christian Ankerstjerne's collection)

Early model Pz.Kpfw. I Ausf. A. with long horn and no additional side armour.

Pz.Kpfw. I Ausf. A (above) and *kleiner Panzerbefehlswagen* (below).

Weihnachten, 1935. Saint Nikolaus, carrying a sack of presents, stands atop a Pz.Kpfw. I Ausf. A with *Buntfarbenanstrich* camouflage. The number of Pz.Kpfw. I reported at this time in the army inventory was 720. This number would grow to 1,065 by 1 June 1936 and to 1,160 by 1 October 1936.

Chapter Three

Pz.Kpfw. I (Sd.Kfz. 101) Ausf. B

The Pz.Kpfw. I Ausf. B was a more powerful tank. Entering service in 1935, it had a larger 100hp engine, giving the tank an increased road speed of 25km/h, with the slightly larger fuel tank giving a cruising range of 170km, or 115km across country.

The Pz.Kpfw. I Ausf. B had a larger water-cooled Maybach six-cylinder NL-38 engine, replacing the previous air-cooled Krupp engine, with an improved six-speed ZF Aphon FG 31 gearbox. With a 90mm bore and 100mm piston stroke, the engine developed 100hp at 3,000rpm.

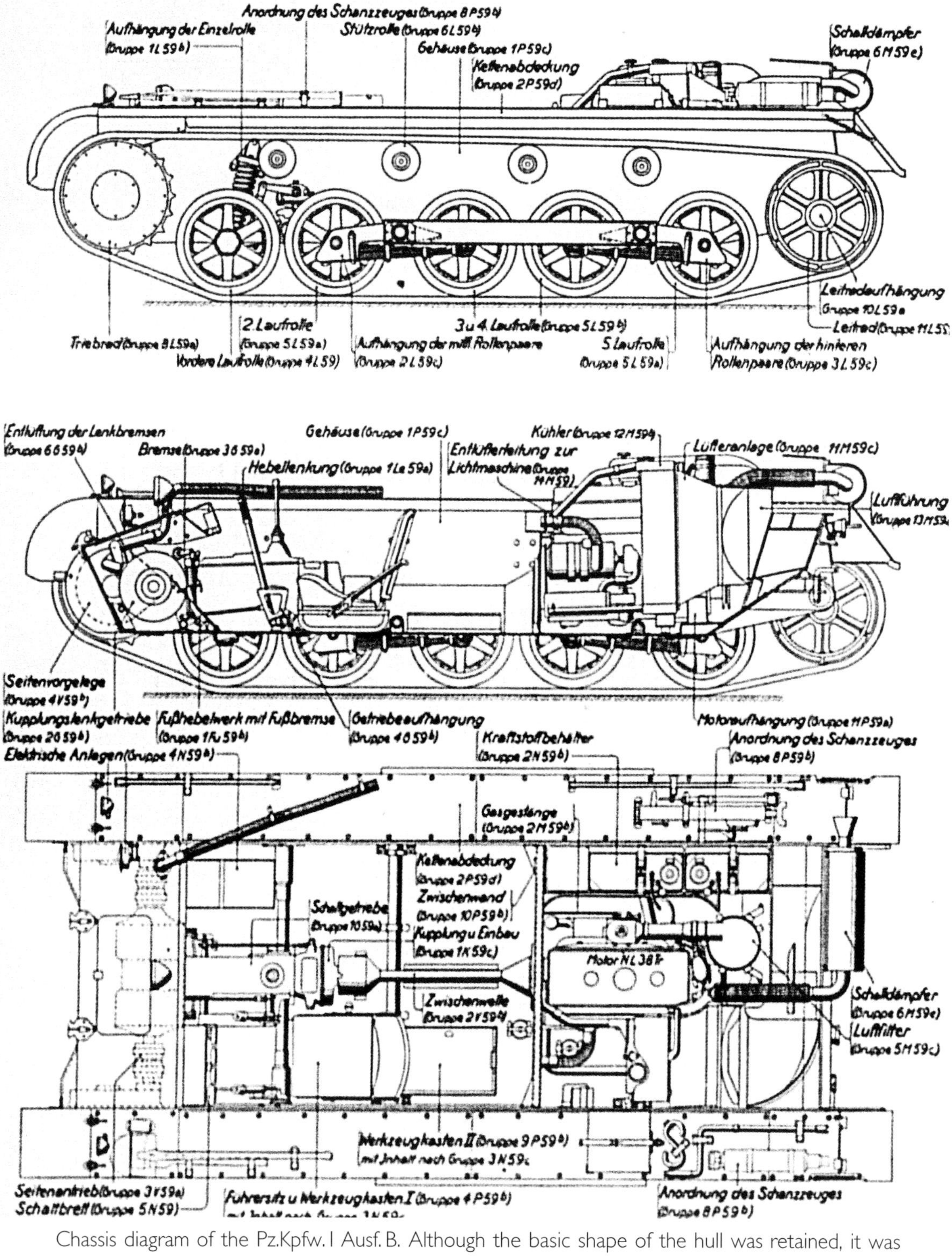

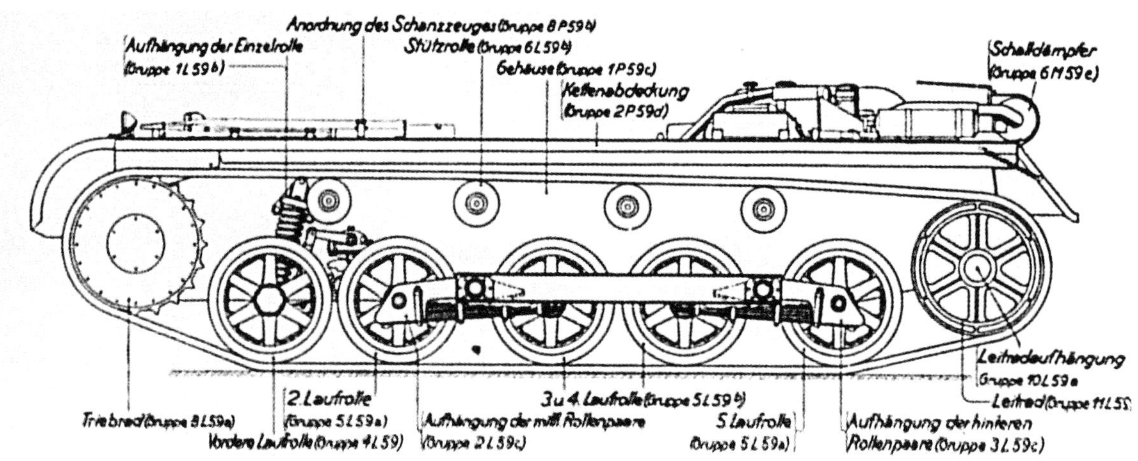

Chassis diagram of the Pz.Kpfw. I Ausf. B. Although the basic shape of the hull was retained, it was extended from 4.02m to 4.42m to accommodate the larger power plant and provide a more stable ride with less propensity to throw a track. The idler wheel was raised and the number of road wheels increased to five with four return rollers. The tank's (combat loaded) weight increased from 5.4 tons to 5.8 tons

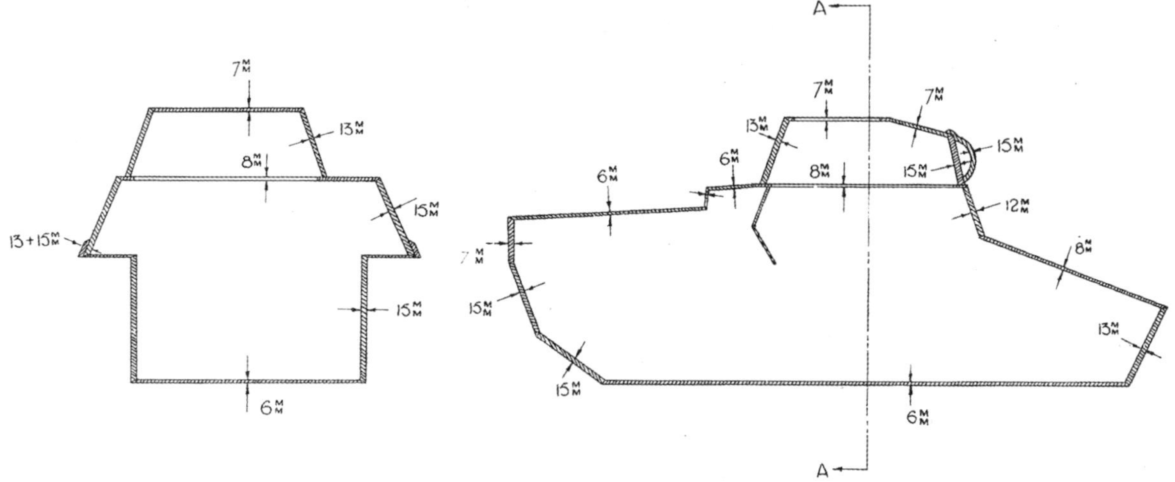

Allied intelligence schematic of the thickness of Pz.Kpfw. I Ausf. B armour.

The Pz.Kpfw. I Ausf. B retained the same armament as the Ausf. A: two MG 13 7.92mm machine guns, the standard German machine gun before the introduction of the MG 34.

Four companies, Henschel, MAN, Krupp-Grusonwerk and Daimler-Benz, built a total of 399 Pz.Kpfw. I Ausf. B. The first examples were completed in July or August 1936.

The rear armour enclosing the engine compartment was redesigned on the Pz.Kpfw. I Ausf. B with a single muffler positioned on the rear right of the vehicle. Another telltale difference was the relocation of the three turret lifting lugs from the sides to the turret roof. Three conical bolts now held the frame for the laminated glass in the vision ports. Note the reinforcing rod on the lower rear hull for the idler housing in the top photograph.

Two images showing the Pz.Kpfw. I Ausf. B on parade in Berlin celebrating Hitler's 48th birthday on 20 April 1937. (NARA)

Maintenance on tanks from the 2nd Panzer Regiment – note the Roman numeral II inside the rhomboid shape on the nearest vehicle. The number '602' corresponds with the 1938 standardization of identification numbers, a system designed to 'ease the command technique, fire control, and disciplinary monitoring'. (Christian Ankerstjerne's collection)

Training at night (top) and in the art of retrieving a bogged vehicle (bottom).

Four images from period postcards showcasing the cross-country capabilities of the Pz.Kpfw. I Ausf. B. The tank could climb a 0.37m vertical obstacle and cross a trench 1.4m wide. In a combat situation, such obstacles would slow down tanks and provide an opportunity to enemy gunners.

Pre-war manoeuvres. The closest tank, an Ausf. B, *Fahrgestell* (chassis) number 14541, was one of the thirty-four examples manufactured by Krupp-Grusonwerk. (Christian Ankerstjerne's collection)

On parade at Berlin's Brandenburg Gate. The Pz.Kpfw. I Ausf. B also differed from its predecessor in having three bolts holding the glass vision block in place on the driver's visor. Former Panzer crewman Armin Böttger recalled shooting practice in a Pz.Kpfw. I: an 'undulating concrete apron had a course marked out with flags. While driving a Panzer Mk I down the course, one had to fire twenty-five rounds at a target the full-size shape of a man. One's view of the outside world was obtained through a periscope. Now one can easily see that because the Panzer rose and fell over the uneven concrete terrain, the barrel of the MG, a fixed installation in the turret, had a field of fire ranging between the concrete and the sky.'

Engine cooling also proved problematic in the Ausf. B under European conditions. Production changes (dated 6 February 1937) incorporated a larger radiator and a more efficient cooling fan.

Inspection. The tube running from the front of the Pz.Kpfw. I Ausf. B provided ventilation to the steering brakes. Note the Tetra fire extinguisher on the top vehicle and the wire cutters now positioned on the right rear track guard. (Galland Books)

Suspension repairs to Pz.Kpfw. I Ausf. B '614' – the number identifying the fourth tank of the first platoon of the sixth company. The three-digit numbering system, introduced in 1938, specified numbers placed on a rhomboid metal plate on the side and rear of a tank. Odd-numbered regiments had white numbers; even-numbered regiments had yellow. Note the reinforcing rod across the lower rear hull. (Galland Books)

Close-up detail on the rhomboid-shaped numbering shield, the two wingnuts allowing the identifying metal plate to be easily removed. (Christian Ankerstjerne's collection)

Close up of the rear-mounted exhaust and perforated cover.

What looks like a joyride in a Panzer-*Schulfahrgestelle* Ausf. B (driving training vehicle), one of 295 manufactured.

Umsetzfahrzeug

Umsetzfahrzeug (converted chassis) tanks featured an expedient training vehicle, created by removing the turret from a standard Ausf. B chassis, retrofitted with an Ausf. A turret. (Galland Books)

A total of 147 *Umsetzfahrzeug* were assembled by Henschel and Krupp-Grusonwerk. Note the tell-tale lifting lugs on the side of the Ausf. A turret.

Chapter Four

Pz.Kpfw. I variants

Pz.Kpfw. I Ausf. C and Ausf. F

The Pz.Kpfw. I Ausf. C (VK6.01) arose from a requirement issued in September 1939 for a new reconnaissance tank. Designated Panzer nA VK6.01 (nA = *neuer Art* or 'new model'; VK = *Vollketten-kraftfahrzeug* − 'fully tracked vehicle', 6 tons, design number 1), the tank was jointly designed by Daimler-Benz and Krauss-Maffei. Weighing 8 tons, the Ausf. C bore little resemblance to its predecessor. The turret carried a 7.9mm *Einbauwaffe* 141 and one MG 34. While the 1st Panzer Division evaluated the tank on the Eastern Front, the majority of the forty examples produced by Krauss-Maffei would serve in a training role.

The two-man (driver, commander/gunner) Ausf. C pictured at Krauss-Maffei circa late 1942 or early 1943. Note the ventilation slits on the engine hatch for deployment in hot conditions. Frontal armour was 30mm thick, side and rear armour up to 20mm.

The tank was 4.195m long, 1.920m wide and 1.945m high. Combat weight was 8 tons.

The Pz.Kpfw. I Ausf. F arose from a December 1939 directive for a new, heavily armoured, infantry support tank. The experimental vehicle co-designed by Daimler-Benz and Krauss Maffei bore the project designation VK18.01.

Like the Ausf. C, the Pz.Kpfw. I Ausf. F (VK 18.01) had interleaved road wheel, torsion bar suspension and a 150hp Maybach HL 45P engine. Protected by a formidable 80mm of frontal armour, it was armed with two MG 34 machine guns. Thirty examples were produced. Eight served in Russia with the 1st Panzer Division with an unknown number assigned to security duties in Yugoslavia. The remainder were used for training.

Thirty Pz.Kpfw. I Ausf. F were produced in 1942. Seven were assigned to the proposed invasion of Malta but instead saw service on the Eastern Front. Production tanks, unlike the prototype, featured radio communication.

Like its Pz.Kpfw. I predecessors, the Ausf. F carried a two-man crew and was armed with dual (MG 34) machine guns. With a combat weight of 21 tons, it had a maximum road speed of 25km/hr and a range of 150km, dropping to 18km/hr and 110 km off road. This tank was serving in Belarus with the 2nd Police Tank Company when it was captured.

kleiner Panzerbefehlswagen (Sd.Kfz. 265)

The *leichte (Funk) Panzerwagen* command tank was based on a Pz.Kpfw. I Ausf. A chassis. Fifteen of these tanks were produced prior to tactical exercises held at Munster in August 1935.

Commander of the 5th Panzer Regiment, *Oberstleutnant* Karl Zuckertort, in a *leichte (Funk) Panzerwagen* command tank, breaks a white tape strung across the Wünsdorf garrison entrance. (NARA)

Pz.Kpfw. I Ausf. A and *kleiner Panzerbefehlswagen* (kl.Pz.Bef.Wg.) on field manoeuvres.

The redesigned superstructure of the kl.Pz.Bef.Wg., seen here in Spanish service, could accommodate an Fu6 radio transmitter and receiver, along with an Fu2 receiver, as well as the three-man crew – commander/gunner, radio operator and driver. (Galland Books)

This kl.Pz.Bef.Wg. has a *Rahmenantenne* (frame antenna) but no commander's cupola. Production commenced in 1936 with 184 vehicles assembled at Daimler-Benz Werk 40 before the end of 1937. Note the strip of additional armour bolted to the superstructure side.

Seen here with a commander's cupola, providing safer observation, this second series kl.Pz.Bef.Wg. was armed with a single MG 34 machine gun in a ball mount on the front superstructure. It was aimed using a *Kugelzielfernrohr* 2 (KZF2) telescopic sight, and some 900 rounds of ammunition (a mix of armour-piercing and tracer) were carried in twelve drum magazines.

Kl.Pz.Bef.Wg. I (Sd.Kfz. 265) with a frame to increase the range. A German report after the French campaign noted that the *Befehlspanzer* was 'too small and has insufficient armour. The commander must have room to move about when he performs his tasks of leading, calling on the radio, firing the machine guns and observing. He also needs the security of sufficient armour.' While in motion it was practically impossible for the driver or commander to reach and adjust a radio set from transmission to reception or to maintain a radio on a fixed frequency without continual adjustment.

Underside of a kl.Pz.Bef.Wg. I lacking a reinforcing bar across the lower rear hull. Powered by a Maybach six-cylinder NL-38 engine, the kl.Pz.Bef.Wg had a cross-country speed of 12–15km/hr and range of 115km. With an average road speed of 25km/hr, the vehicle had a range of 170km. It was capable of climbing a 30° incline.

4.7cm Pak (t) (Sfl.) *auf Panzerkampfwagen I ohne Turm* (Sd.Kfz. 101)

Originally designated 4.7cm Pak (t) (Sfl.) *auf Panzerkampfwagen I ohne Turm*, the *Panzerjäger* I was a self-propelled gun that merged a modified Pz.Kpfw. I Ausf. B chassis with a Skoda 4.7cm PaK 36(t) L/43.4 anti-tank gun. It was conceived to combat heavy French tanks. A prototype vehicle was demonstrated before Hitler and senior army representatives on 10 February 1940.

In total, 202 vehicles were converted in 1940–1941, 132 of them by Alkett and a second series of 70 vehicles by Alkett and Klöckner-Humboldt-Deutz AG.

The second series *Panzerjäger* I featured a seven-sided shield. Neither version had rear or overhead protection, and it was subsequently reported that an enemy soldier could kill the crew by hurling a grenade from a foxhole 'close to the side or rear of the *Panzerjäger*'. The commanding officer of Panzerjäger Abteilung 643 recorded that observation was particularly bad, leaving the crew effectively blind 'when attacking in villages, or against street barricades, machine-gun nests and individual tanks'.

Panzerjäger I crew, mostly attached to Panzer Divisions, wore the black Panzer uniform. Helmets provided protection inside the open compartment.

15cm-*schweres Infanteriegeschütz* 33 (Sf) *auf Pz.Kpfw. I ohne Aufbau Ausf. B (Sd.Kfz. 101)*

The need for mobile armoured artillery, realized during the Polish campaign, led to the conversion by Alkett of thirty-eight of these self-propelled guns in February–March 1940. The upper hull of a Pz.Kpfw. I Ausf. B was removed and replaced with a 15cm *schwere Infanteriegeschutz* (heavy infantry gun) 33 gun on its field carriage within a new box-shaped armoured superstructure. The original chassis remained unchanged, leading to frequent breakdowns from the failure of brakes, clutches or tracks. Nevertheless, a report from September 1941 stated it had performed 'very well as an assault gun' during the invasion of Russia.

Miscellaneous variants ...

One of several types of Pz.Kpfw. I Brückenleger (bridge-layer). An unknown number of vehicles were produced with service in Poland, France and the Soviet Union. (Collection Akira Takiguchi)

Whitewashed 2cm FlaK 38 *auf* Pz.Kpfw. I Ausf. A, a *Flakpanzer* built using obsolete tank chassis in 1941. Some twenty-four examples were produced.

Sanitätskraftwagen I *auf* Pz.Kpfw. I Ausf. B, an open-top ambulance variant from the 1st Panzer Division, crosses a pontoon bridge over the Meuse river at Floing, near Sedan. (Galland Books)

The *Munitionsschlepper auf* Pz.Kpfw. I Ausf. A was a basic conversion with the turret removed and replaced by a two-piece hatch.

An unarmed, turretless *Instandsetzungspanzer* I passes a Pz.Kpfw. III in Bulgaria, March 1941. This service variant carried spare parts and tools for small repairs and maintenance. Note the multiple *Kanister* carried on the rear.

The 3.7cm PaK 35/36 *auf Fahrgestell* Pz.Kpfw. I Ausf. A, likely a field modification incorporating an anti-tank gun on a redundant chassis.

This instructional oddity featured a wooden superstructure and Pz.Kpfw. III turret on a Pz.Kpfw. I Ausf. A chassis. It was powered by a *Holzgasgenerator* (wood gas generator). See also page 288. (Collection Akira Takiguchi)

Alkett's *Versuchs Kraftfahrzeug* 617 (Vs.Kfz. 617) *Minenräumer* was a prototype vehicle intended to explode mines using the heavy 'shoes' on its wheels. Armed with a Pz.Kpfw. I turret with two MG 34 machine guns, the ponderous 55-ton vehicle was easily bogged during testing. Manufacturing plans were shelved. Captured by Russians forces in 1945, the Vs.Kfz. 617 is today displayed at the Kubinka tank museum.

Chapter Five

Pz.Kpfw. I on film

Tag der Freiheit

German film director Helene 'Leni' Riefenstahl's third propaganda film, *Tag der Freiheit: Unsere Wehrmacht* ('Day of Freedom: Our Armed Forces'), was the first public display of Germany's new *Panzerwaffe* in mock combat before Hitler and an audience of thousands of spectators.

Released theatrically on 30 December 1935, *Tag der Freiheit* helped promote Nazi Germany's burgeoning military to the world.

Approaching a camera at speed, this Pz.Kpfw. I Ausf. A was filmed in Nürnberg, the city of the *Reichsparteitage* (Reich Party Congresses), during the seventh NSDAP Party Congress in 1935. Riefenstahl used the film to celebrate the rearmament of Nazi Germany's army.

Verräter

The premise of the 1936 Nazi propaganda film *Verräter* ('Traitor') involved a British spy ring attempting to uncover the secrets of the German defence industry. Spies infiltrate the Heinkel aircraft factory and approach an inventor working in the fictitious T-Metallwerke department, where the latest German tank was being developed. They also approach a former bank clerk, Hans Klemm, now a Pz.Kpfw. I gunner, who struggles to report their activity to his superiors, who in turn, notify the Gestapo. The 92-minute film was commissioned by Admiral Wilhelm Canaris, head of military intelligence (Abwehr). Director Karl Ritter's feature film showcased the Third Reich's Luftwaffe, fledgling *Panzerwaffe* and secret police. It starred matinée idol Willy Birgel and Joseph Goebbels' mistress Lida Baarova. *Verräter* received a preview screening on 9 September 1936 at the Nürnberg Party Congress, before premiering six days later in Berlin at the Ufa Palast am Zoo. Representing Germany, the film was the main entry at the 1936 4th Venice International Film Festival, where it received a special commendation.

Directed by Karl Ritter, *Verräter* introduced to a global audience the Pz.Kpfw. I Ausf. A as Nazi Germany's cutting-edge tank. The film premiered in the US in New York on 27 January 1937. *Variety* (27 January 1937) praised the production, writing: 'Verräter is in every aspect a class A piece of celluloid, if the matter of political viewpoints is left aside.'

Verräter was filmed at Wünsdorf using tanks from the 5th Panzer Regiment.

(**Opposite, above**) Passing through a forested area (with turrets reversed), the tanks race down a hill and at Zossen open fire on silhouetted-soldier targets. *Verräter* provided both a warning to foreign spy agencies and a message to ordinary Germans. *New York Times* film critic Claire Trask described the film as 'super-educational', characterizing a 'traitor' and, by association, a 'patriot'. The Propaganda Ministry published a 32-page study guide for high schools, while the film was also shown to workers in heavy industry to encourage them to be alert for foreign espionage.

(**Opposite, below**) Pz.Kpfw. I Ausf. A race towards the camera.

Synchronized hand signals were portrayed in the film as the method of communication.

Verräter ends with an announcement that the traitor Fritz Brockau (Rudolf Fernau) will be executed. The character Hans Klemm (Heinz Welzel), a Pz.Kpfw. I commander, is congratulated for his courage in front of the assembled Panzer regiment for helping to apprehend Brockau.

The film closed with scenes of massed Panzer and bomber formations.

Note the unusual Pz.Kpfw. I command vehicle in a frame taken from the movie.

Scenes from the German propaganda film *Feldzug in Polen* (Campaign in Poland) featuring a Pz.Kpfw. I Ausf. B and destroyed Ausf. A. The film premiered in the Berlin Ufa-Palast um Zoo before a theatrical release in February 1940.

Sieg im Westen

The campaign in France was the subject of the feature-length 1941 film *Sieg im Westen* ('Victory in the West'). General Erwin Rommel was asked to participate, both in front and behind the camera, using the 7th Panzer Division to recreate the crossing of the Somme, a role for which he was later uncredited.

Stills from *Sieg im Westen*: a Pz.Kpfw. I Ausf. B (with a single machine gun) emerges from a camouflaged position before a staged attack on the Chateau le Quesnoy.

Further stills from *Sieg im Westen*: Pz.Kpfw. I Ausf. B shown during the re-enactment of the storming of Hangest-sur-Somme, the scene of bitter fighting in June 1940. During the original battle Rommel directed a Panzer battalion to 'shoot up the enemy' on the western side of the village without entering it.

Pz.Kpfw. I Ausf. B advance as Hangest is 'retaken'.

A 15cm s.I.G. 33 (Sf) auf Pz.Kpfw. I Ausf. B enters Hangest, before firing on a building at point-blank range. (NARA)

Kampf um Norwegen – Feldzug 1940

Newsreel footage from the propaganda film *Kampf um Norwegen – Feldzug 1940* ('Battle for Norway – 1940 Campaign') showing a Pz.Kpfw. I Ausf. A negotiating a rocky pass. The tanks bears the later style *Balkenkreuz* (bar cross), introduced on 26 October 1939.

Chapter Six

Pz.Kpfw. I in foreign service

Krupp's *Leichter Kampfwagen Ausland* (LKA 1) was a prototype export light tank weighing 4.5 tonnes. Influenced by the design of the Pz.Kpfw. I Ausf. A, the tank's Krupp V8 air-cooled engine delivered 85hp and a top speed of 50km/h. A restriction on armament by Wa. Prüf. 6 (in addition to a host of caveats such the type of bolts and vision slits used) would force any buyer to use foreign weaponry. A later prototype mounted a 2cm KwK 30 gun. Although a number of potential buyers showed interest in the tank, no orders were forthcoming.

Krupp's *Leichter Kampfwagen B* (LKB 1), another proposed export model, was fitted with an air-cooled Krupp M601 diesel engine developing only 45hp. The trials proved unsuccessful and petrol engines were used in all subsequent German armoured vehicles except for the Sd.Kfz. 234 8 × 8 armoured car, which was powered by a Tatra 103, V-12 air-cooled diesel engine. Note the angled side exhaust on the remodelled rear deck. A later model, the LKB 2, was similar to the Pz.Kpfw. I Ausf. B with five road wheels and a raised idler. While records exist of anticipated LKB sales to Turkey, Afghanistan, Peru, Sweden and Bulgaria, none was exported.

China

Former Chinese Pz.Kpfw. I Ausf. A captured by Japanese forces during the fighting for Nanjing in December 1937 during the Second Sino-Japanese War. Fifteen of these tanks were sold to the Republic of China, arriving in mid-1937. All were captured by Japanese forces. A Krupp representative sent to China reported on the need for all hatches and visors to be opened during the summer months. The heat would be unbearable after a long march, and the gunner would be unable to operate the machine guns with hatches and visor closed.

Former Chinese Pz.Kpfw.I Ausf. A captured in Nanjing. The export version featured simplified *Sehklappen* (vision ports). A report from China stated that the tanks were inadequately prepared for the sea voyage, arriving with the gunsights, machine-gun mounts and steering brakes heavily corroded.

Former Chinese Pz.Kpfw. I Ausf. A on parade (right) and display at Yasukuni-jinja, a Shinto shrine in Tokyo commemorating Japan's war dead, January 1939.

Trophies of war displayed at Japan's Hanshin Koshien Stadium in February 1939: Renault FT, Pz.Kpfw. I Ausf. A, T-26 and Vickers 6-ton Type B. Chinese authorities criticized the armament of the German tank as too weak.

Spain

War broke out in Spain on 17 July 1936 in the aftermath of a failed *coup d'état* by a group of generals against the Popular Front government. In their fight against leftist Republican forces (championed by the Soviet Union), *Generalissimo* Francisco Franco's Nationalist forces received German tanks and training personnel under the leadership of *Oberstleutnant* Wilhelm Ritter von Thoma. In Germany the 6th Panzer Regiment assembled on 20 September 1936 after a call for volunteers for a secret mission abroad under combat conditions. Nearly all the men stepped forward, unaware they would be serving in Spain. Further volunteers from the 3rd, 4th and 5th Panzer Regiments followed. A total of some 102 Pz.Kpfw. I would eventually be shipped to the Nationalists, roughly half of which were Ausf. A and the rest Ausf. B.

(**Opposite, above**) Under a new flag in *Buntfarbenanstrich* camouflage. On 7 October 1936 an initial shipment of forty-one Pz.Kpfw. I Ausf. A arrived in Spain. A further twenty-one Ausf. B followed, with another ten replacements arriving in early 1937. A training camp was set up in Santa Juana de la Cruz de Cubas de la Sagra, outside Madrid, with German volunteers supporting the establishment of two Spanish tank battalions. (Galland Books)

Camouflaged Pz.Kpfw. I Ausf. A with its Spanish crew. Note the covered machine-gun barrels and fine branches used as camouflage. The Spaniards thought highly of the MG 13. (Galland Books)

German Pz.Kpfw. I Ausf. A in service with Franco's Nationalist forces. Rather than Spain serving as a 'proving ground' for Hitler's first light tank, in fact the Pz.Kpfw. I was never deployed en masse with other arms in Spain as it would be in the forthcoming European campaigns. Combat reports from Spain revealed that the seal for the gun mantlet was insufficient, with bullet fragments able to pierce the gap between the mantlet and turret, wounding the gunner.

Unarmed Pz.Kpfw.I from *Panzerabteilung Thoma* with the national Spanish flag on the front superstructure. Practical experience in Spain revealed a flaw in the turret seal, which let in rainwater, rusting components. A Molotov cocktail would prove disastrous.

Spanish-crewed Pz.Kpfw.I Ausf. A. Colonel Wilhelm Ritter von Thoma, commander of German forces in Spain, reported that the German tanks achieved initial success in close combat using armour-piercing (S.m.K.) bullets against Soviet-manufactured T-26 tanks before the Republicans learned the benefit of distance and their superior 45mm gun. Note the *Tercio* emblem painted beside the Spanish flag. (NAC)

A Pz.Kpfw. I Ausf. A and captured Soviet T-26 leave Esplugues de Llobregat on 26 January 1939 in preparation to enter Barcelona.

Rather than Spain serving as a 'proving ground' for Hitler's first light tank, in fact the Pz.Kpfw. I was never deployed en masse with other arms in Spain as it would be in the forthcoming European campaigns. Combat reports from Spain revealed that the seal for the gun mantlet was insufficient, with bullet fragments able to pierce the gap between the mantlet and turret, wounding the gunner.

Direct hit: a Pz.Kpfw. I Ausf. A turret blown from the superstructure of a Nationalist tank. A contemporary German report noted that turrets had been jammed by rifle fire, though they had also 'held up to hits by small shells with sensitive fuses'. (Galland Books)

This Pz.Kpfw. I Ausf. A was knocked out by Republican forces outside Madrid in November 1936. A period report noted that the 'crew are totally helpless when they can't traverse the turret, such as when the tank is nose down in a large shell hole'. A suggestion was forwarded for pistol ports to be incorporated into the tank's visors. (Galland Books)

Spanish-crewed kl.Pz.Bef.Wg. Production delays were responsible for the lack of a ball mounting for the single MG 34 machine gun. Four of these command vehicles were sent to Spain. (Galland Books)

Spanish Nationalist forces entered Bilbao on 19 June 1937 after a seven-day battle. The Teatro Arriaga (*teatro arriaga antzokia* in Basque) opera house in the background was designed in neo-baroque style by architect Joaquín Rucoba and completed in 1890. (Galland Books (bottom))

Jubilant locals atop a Pz.Kpfw. I Ausf. A in Bilbao. A handful of captured examples were displayed by the Republicans in Barcelona. Reports from Allied commentators ridiculed the tank: a French observer believed the 'German Panzer division has failed before it was even put into service', with French models remaining the 'King of the Battlefield'; and British general J.F.C. Fuller likened the cramped vehicle to a 'mobile coffin', with the war in Spain demonstrating how 'the light tank is absolutely no combat machine'.

Republican forces enter Santander on 26 August 1937. (NARA)

To increase the firepower of the Pz.Kpfw. I, four tanks were upgraded with an Italian Breda Model 1935 20mm gun in a modified turret. Spanish authorities also pressed Germany to send tanks 'if possible with a 20mm cannon or larger', but to no avail.

Two Pz.Kpfw.I were modified by Spanish Nationalists as flamethrowing tanks, unofficially dubbed Pz.Kpfw.I *Lanzallamas*, by replacing the machine guns with the *Flammenwerfer* 35. This Ausf.A *Lanzallamas* is shown during testing at Cubas in 1937. Both German and Italian flamethrower units were fitted and tried. Neither saw combat.

The *Panzertruppenabzeichen der Legion Condor* (Condor Legion Tank Badge) was inaugurated in September 1936 by Colonel von Thoma to recognize German tank crews who had served at least three months in Spain. A total of 416 were awarded.

Other foreign service examples

Four Pz.Kpfw. I Ausf. A were used by the Independent State of Croatia (NDH), a German puppet state established on 10 April 1941. A single Pz.Kpfw. I Ausf. B was sold to Hungary in 1938.

Chapter Seven

Pz.Kpfw. I and variants at war

Pz.Kpfw.I Ausf.A and Pz.Kpfw.II line-up in Karlsbad, Sudetenland, with the regimental flag, on 4 October 1938. In April 1939 a British MP, amid 'loud laughter' in the House of Commons, asked Prime Minister Neville Chamberlain 'What proportion of German army tanks are made of plywood?' Allegedly, a British motorist 'cruising along a lane in Germany turned a corner and crashed into a tank. It fell to pieces.' Years after the Reichswehr paper panzers, Hitler's Pz.Kpfw.I had already seen combat in Spain and was months away from crossing the Polish frontier. (BA 146-2005-0178)

Oberbefehlshaber des Heeres (Commander of the Army) Generaloberst Walter von Brauchitsch reviews a Pz.Kpfw. I Ausf. B column entering the Sudetenland on 13 October 1938. From July 1937 tanks were painted in *Dunkelgrau* (dark grey) with splashes of *Dunkelbraun* (dark brown). Directives called for the paint to be applied using a spray gun (for less shine) with feathered edges between the colours. (BA146-2005-0181)

Fall Weiß – the invasion of Poland

(**Opposite, above**) *Umsetzfahrzeug*, Pz.Kpfw. II and Sd.Kfz. 251/3 pictured during the invasion of Poland. The German Army assigned 973 Pz.Kpfw. I to the campaign – nearly 40 per cent of the 2,553 tanks involved – with 260 held in reserve. Note the slotted, cylindrical, short Bosch horn. German propaganda explained the rationale for invading Poland: 'Why are we fighting? Because we were forced into it by England and its Polish friends. If the enemy had not started the fight now, they would have within two or three years. England and France began the war in 1939 in fear because in two or three years, Germany would be militarily stronger and harder to defeat.' (BA 146-1976-071-36)

(**Opposite, below**) Pz.Kpfw. I Ausf. A (no. 425) and Pz.Kpfw. II from the 1st SS Panzer Division Leibstandarte SS Adolf Hitler crossing the Bzura river in central Poland. Reports were spread by Axis propaganda of Polish cavalry attacking German tanks with lances. (Galland Books)

Pz.Kpfw. I Ausf. A crossing a pontoon bridge. The solid white *Balkenkreuz* on the turret sides followed a directive issued on 13 July 1939 for the symbol to be painted on all four sides of a tank. Armoured vehicles not displaying it were to be considered the enemy. (Galland Books)

Camouflaged Pz.Kpfw. I Ausf. A on the banks of the Brda river in northern Poland. Despite not having the capacity to destroy enemy armour, the tank was nevertheless effective against soft-skinned targets. (BA 146-1978-120-11)

Pz.Kpfw. I Ausf. B '33' was immobilized by a Polish anti-tank mine in northern Pless (present-day Pszczyna). Note the retrofitted *Nebelkerzenabwurfvorrichtung* (NKAV) grenade launcher. These photos were taken on 5 September 1939 after the fighting at Pszczyna, in which Germany lost five tanks. (NARA)

Close-up of the NKAV grenade launcher, a device that discharged *Schnellnebelkerze* 39 smoke grenades. It was typically fitted on most German tanks from 1939 to 1942. The device was activated by a cable from inside the tank, releasing one grenade at a time. (BA 101I-163-0332)

An early Pz.Kpfw. I Ausf. B and Pz.Kpfw. II similarly halted by Polish anti-tank mines at Pszczyna. (NARA)

Pz.Kpfw. I. Ausf. A with rails to carry and deploy fascines; a Pz.Kpfw. II Ausf. B in the foreground.

Polish prisoners of war carried on a Pz.Kpfw. I Ausf. A. from the 4th Panzer Division.

Pz.Kpfw. I Ausf. B with NKAV. Note the reversed track on '313' (has he noticed?) and the less conspicuous Balkenkreuz on the vehicles in the lower photo. (BA 101I-012-0035-11A (bottom))

A scene during the attack on the Vistula, with German tanks from the 4th Panzer Division capturing Polish supply vehicles near Wyszogród. The Pz.Kpfw. I Ausf. B closest to the camera is possibly from a reconnaissance platoon.

Pz.Kpfw. I Ausf. B '104' south of Wyszogród during the attack on the Vistula, 18 September 1939.

Pz.Kpfw. I Ausf. B with muted *Balkenkreuz*. A directive issued on 12 October 1939 recognized the conspicuous white crosses as 'extraordinarily dangerous … as they can be successfully used by enemy anti-tank defences as aiming points'. As a consequence, the *Balkenkreuz* was to be 'removed IMMEDIATELY' and an alternative recognition method such as flags and pennants used instead. Note the armour reinforcing strip bolted to the side of the superstructure. (Galland Books)

Pz.Kpfw. I Ausf. B pictured in Przemysl. The Cathedral Basilica of the Assumption of the Blessed Virgin Mary and St John the Baptist rises in the background.

Repairs to the suspension of an *Umsetzfahrzeuge*. Note the Panzer crewmember in the foreground wearing a standard army *Feldmütze* (side cap). (Galland Books)

Pz.Kpfw. I Ausf. A from the 4th Panzer Division featuring the formation's distinctive yellow Balkenkreuz with white outline. (Christian Ankerstjerne's collection)

A weathered Pz.Kpfw. I Ausf. B, minus its left track, in Warsaw. In total, eighty-nine Pz.Kpfw. I were written off in Poland, to be either scrapped or comprehensively overhauled. (Galland Books)

Inquisitive soldiers inspect a Pz.Kpfw. I Ausf. A at the Saxon Gardens in Warsaw, one of Poland's oldest public parks.

An unarmed kl.Pz.Bef.Wg. serving as an armoured ambulance. (Galland Books)

Die von der Panzerkompanie

Worte: Gefreiter Hans Linden.
Weiſe: Soldaten W. Becker und W. Tolksdorf.

Im Herbſt die Vögel flogen gen Süden übers Land,
Soldaten aber zogen in Fernen unbekannt.
Hei, hei, fröhlich waren ſie, die von der Panzerkompanie;
hei, hei, fröhlich waren ſie, die von der Panzerkompanie.

Wir zogen ins Sudetenland; Sudetenland ward frei.
Ein Jubel war es ungekannt nach aller Tyrannei.
Hei, hei, fröhlich waren ſie, die von der Panzerkompanie uſw.

Gen Prag ging es in Schnee und Eis, es war ein hartes Stück;
und kehrten wir von dieſer Reiſ' zu Oſtern nicht zurück:
Hei, hei, fröhlich war'n doch ſie, die von der Panzerkompanie uſw.

Als wieder Herbſt im weiten Feld, lag Polens Macht und Wehr
beſiegt, geſchlagen und zerſpellt, war's auch mitunter ſchwer.
Hei, hei, fröhlich kämpften ſie, die von der Panzerkompanie uſw.

Und weiter tun wir unſre Pflicht, in Frieden oder Krieg;
ſelbſt wenn des Letzten Auge bricht, uns Panzern iſt der Sieg.
Hei, hei, fröhlich bleiben ſie, die von der Panzerkompanie uſw.

Those from the Panzer Company
Words: Corporal Hans Linden.
Tune: Soldiers W. Beder and W. Tolksdorf.

In the autumn the birds flew south over the land, but soldiers moved into unknown remote regions. Hey, hey, they were happy, those from the Panzer Company; hey, hey, they were happy, those from the Panzer Company.

We went to the Sudetenland; Sudetenland became free. It was a jubilation unprecedented after all the tyranny. Hey, hey, they were happy, those from the Panzer Company, etc.

We went towards Prague in snow and ice, it was a tough thing; and we didn't return from this journey at Easter. Hey, hey, they were happy, those from the Panzer Company, etc.

When autumn was back in the wide fields, Poland's power and defence lay broken, defeated and shattered, it was also difficult sometimes. Hey, hey, they fought happily, those of the Panzer company, etc.

And we continue to do our duty, in peace or war; even if the life's eye of the last breaks, the victory belongs to us Panzers. Hey, hey, they remain cheerful, those of the Panzer company, etc.

Hitler and his entourage (including *General* Heinz Guderian, *Feldmarschall* Wilhelm Keitel and *Reichsführer* Heinrich Himmler) in Poland watching what is most likely a Pz.Kpfw. I Ausf. A trundling past. Note the turret hook, one of three on the turret used for lifting the turret or attaching camouflage material.

Weserübung – the occupation of Denmark and Norway

Pz.Kpfw. I Ausf. A and kl.Pz.Bef.Wg. in Åbenrå, Denmark, on 9 April 1940. Panzer-Abteilung z.b.V. 40 was equipped with twenty-nine Pz.Kpfw. I for the invasion of Denmark and Norway. Note the symbol denoting the 1st Company, Panzer-Abteilung z.b.V. 40.

Locals examine *Umsetzfahrzeug* '632', which came to grief on *Hovedvej* 10 (Highway 10), north of Haderslev, when it stuck a tree, damaging the right drive sprocket. (Christian Ankerstjerne's collection)

'Supplying tanks by sea' reads the contemporary caption. 'The exemplary cooperation of branches of the military made the bold Norwegian campaign a success. Alongside the navy, which itself undertook transport and security at sea, the merchant navy played a commendable role in resolving the difficult supply problem.'

Norwegian civilians on a train platform beside freight cars carrying Pz.Kpfw. I Ausf. B. Rail transport was used where available to limit the mechanical strain on vehicles.

Infantry, dismounted from their bicycles, shelter behind Pz.Kpfw. I Ausf. A. '13' during a clash with Norwegian mountain troops on 21 April 1940. (NARA)

Pz.Kpfw. I Ausf. A '13' pushes forward in Norway.

Oberstleutnant Ernst Volckheim, veteran of the German armoured attack on Villers-Bretonneux in 1918 and now commander of Panzer-Abteilung z.b.V. 40, salutes passing kl.Pz.Bef.Wg. outside the University of Oslo on 1 October 1940.

Fall Gelb – the attack on Luxembourg, the Netherlands and Belgium

German armour crosses the Vroenhoven bridge, one of two bridges captured intact over the Albert Canal, during the invasion of Belgium, the Netherlands and Luxembourg. Note the *Balkenkreuz* applied to the rear of the Pz.Kpfw. I Ausf. A in the foreground. (BA 146-1985-038-06)

'The Panzer Division', this photo was originally captioned, 'pushes into Holland by surprise along field paths.' Note the armoured cover on the rear deck of the Pz.Kpfw. I Ausf. A.

Pz.Kpfw. I Ausf. B follow Pz.Kpfw. II through Valkenburg, in the Dutch province of Limburg, on the morning of 10 May 1940. Forty-three German were killed on this day in the fighting to take the village and the nearby airfield; the Dutch lost fifty-four men. (NARA)

An army NCO trains his 8mm camera on his surroundings as he advances into Holland in an *Umsetzfahrzeug* belonging to the 4th Panzer Division. (NARA)

An Ausf. B commander from the 9th Panzer Division in conversation with a German soldier on a Rotterdam street following the surrender of Dutch forces on 14 May 1940. Instructions issued on 1 November 1939 called for divisional insignia on tanks to feature (in yellow) on the front and rear of the vehicle, as well as on the left superstructure.

Repairs to broken down tanks, an *Umsetzfahrzeug* and Pz.Kpfw. IV, at Bertrix, Belgium, on 12 May 1940. Note the 1st Panzer Division symbol on the rear of the *Umsetzfahrzeug* turret and its chocked tracks. (NARA)

The 4th Panzer Division during the battle of Gembloux on 14–15 May 1940. Pz.Kpfw. I Ausf. A '153' is equipped with an Fu 2 radio receiver. Note the antenna and protective wooden housing which swung onto a horizontal position as the turret traversed to the right. *Grossen Drahtschere* (large wire cutters) are positioned beside the 3-ton jack. (Galland Books)

France

Pz.Kpfw. I Ausf. B leading a Pz.Kpfw. 38(t) column. The number of Pz.Kpfw. I in front-line service had decreased with the availability of other tanks such as the Czech and Pz.Kpfw. III, with 554 available at the beginning of the campaign.

Engelfontaine, northern France, 21 May 1940. Pz.Kpfw. I Ausf. B '822' trains its guns right onto a target. Note the tank's lowered 2m rod antenna and wooden stowage tray.

Pz.Kpfw. I. Ausf. A number 144 from the 7th Panzer Regiment (outlined bison symbol), 10th Panzer Division.

French colonial prisoner of war on a *Umsetzfahrzeug* from General Erwin Rommel's 7th Panzer Division. Nearly 100,000 French colonial soldiers were captured by Germany during the war.

Pz.Kpfw. I Ausf. A in action. Note the armoured covers on the engine cooling vents and *Nebelkerzenabwurfvorrichtung*.

Panzerjäger I (first series). Four Panzerjäger Abteilung (521, 616, 643 and 670) participated in the French Campaign, where they quickly proved themselves in battle. According to a report from the 18th Infantry Division, the vehicle was 'very effective against tanks and also against houses when fighting in towns'.

15cm s.I.G. 33 (Sf) auf Pz.Kpfw. I Ausf. B from the 7th Panzer Division's *schwere Infanteriegeschütz-Kompanie (mot.S)* 705, in action during the Battle of France, 1940. (Galland Books)

The *Abwurfvorrichtungen auf* Pz.Kpfw. I Ausf. B was a demolition variant that could deliver an explosive charge of up to 50kg. Both of these scenes are in France in 1940; note the French Renault D1 in the top photograph.

Abwurfvorrichtungen auf Pz.Kpfw. I Ausf. B beside a Pz.Kpfw. II (likely an Ausf. C).

Abwurfvorrichtungen auf Pz.Kpfw. I Ausf. B with an alternative pattern charge delivery system. A small number of these vehicles were later fitted with a turret flamethrower during the invasion of Russia.
(Galland Books)

Abwurfvorrichtungen auf Pz.Kpfw. I Ausf. B during the occupation of northern France.

A second Pz.Kpfw. I Ausf. B laden with personal effects.

Pz.Kpfw. I Ausf. B '223' from the 7th Panzer Division was hit by Allied fire during the fighting around Arras in May 1940. Dispensing with the earlier metal shield, numbers were painted directly onto the turret.

Kl.Pz.Bef.Wg. R02, identified as the second tank of the 5th Panzer Regiment headquarters, was knocked out on 13 June 1940. The occupants, *Oberfunkmeister* Fritz Heister, *Gefreiter* Otto Hahn and *Gefreiter* Werner Borgmann were all killed.

The three-man crew of this *Panzerjäger* I, killed during the Battle of France, lie buried beside their immobilized vehicle.

Wounded British troops transported on a Pz.Kpfw. I Ausf. B after the fall of Calais in May 1940. (BA 183-B14898)

Pz.Kpfw. I Ausf. A from the 3rd Panzer Division in Dijon, which the division passed through on its return to Germany. Note the French helmet on the glacis plate. A directive issued on 31 July 1940 stipulated that vehicles were to be painted in a single shade of *Dunkelgrau*. (Galland Books)

Externally the *Panzerjäger* I differed in the design of the fixed protective shield around the gun. Vehicles produced in the first series manufactured by Alkett had a five-sided shield of armour 14–15mm in thickness. (Galland Books)

(**Above**) Immobilised Pz.Kpfw. I Ausf. A photographed during the second phase of the campaign in France starting in June 1940. Note the two dots on the hull side denoting the 2nd Panzer Division. During the first phase of *Fall Gelb*, the division was assigned to Panzergruppe von Kleist and carried a large 'K' symbol. During the second phase of the campaign, the division was part of Panzergruppe Guderian with a prominent 'G' symbol. (Galland Books)

(**Opposite, above**) Paris parade: a Pz.Kpfw. I Ausf. B from the Leibstandarte SS Adolf Hitler on the Champs-Elysees on 29 July 1942 before *Generalfeldmarschall* Gerd von Rundstedt. (BA 101I-256-1234-06)

(**Opposite, below**) Landing trials under way, with a Pz.Kpfw. I Ausf. B and Panzerjäger I rehearsing for Operation *Seelöwe* (Sea Lion), the planned invasion of the south coast of the United Kingdom. Some 1,900 river barges were requisitioned for the invasion, with an elementary bow landing ramp installed.

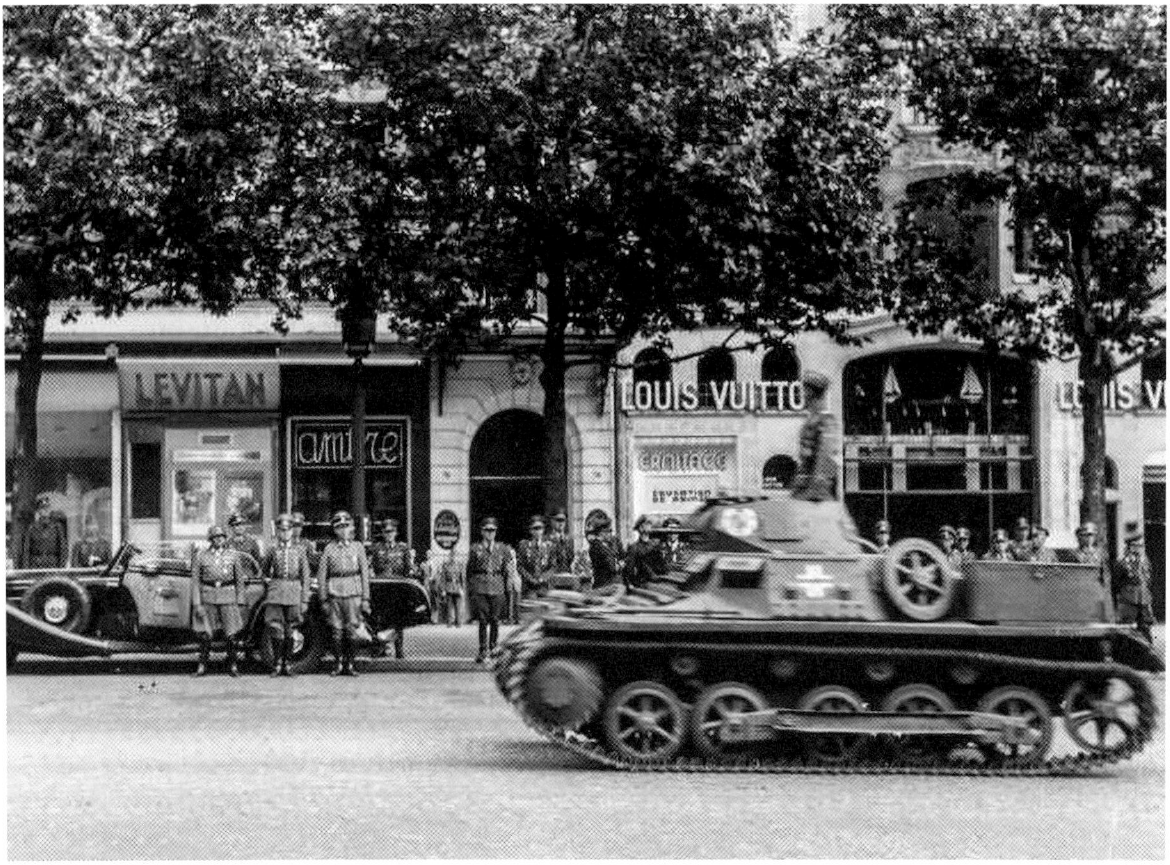
LEVITAN
LOUIS VUITTO
LOUIS VU

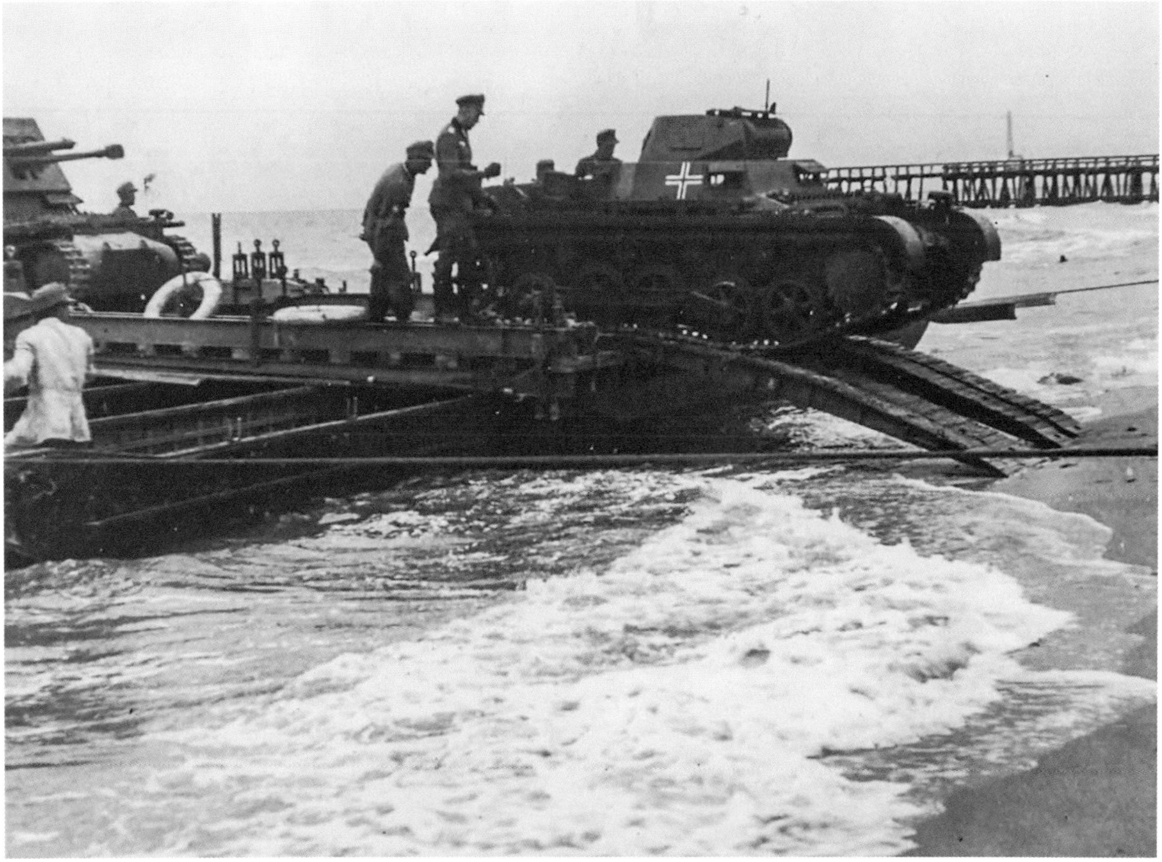

Marita – the invasion of Greece

Pz.Bef.Wg. from the 11th Panzer Division with later pattern commander's cupola incorporating a round signal port in the hatch. The 5.88-ton vehicle could ford 60cm of water.

15cm s.I.G. 33 (Sf) auf Pz.Kpfw. I Ausf. B from the 5th Panzer Division. In a 1945 report, the US Office of the Chief of Ordnance reported: 'This equipment consists of the 15cm heavy infantry howitzer mounted on a turretless Pz.Kpfw. I model "B" chassis … Due to the additional weight carried by the chassis, which approximates 3 tons more than its normal Pz.Kpfw. I complement, its road performance does not equal that of the Pz.Kpfw. I tank. The general appearance suggests that the equipment is overloaded. The howitzer, which probably retains its wheels and trails, in addition to its original shield, traversing and elevating mechanisms, is mounted high in a tall, three-sided shield, and fires forward. The shield is 10mm thick and is open at the top and rear.' (BA 1011-163-0328-15)

Abwurfvorrichtungen auf Pz.Kpfw. I Ausf. B from the 2nd Panzer Division advancing on the road from the Tembi Valley to Larissa in Greece, 19 April 1941. Only nine Pz.Kpfw. I were still operational among the six Panzer divisions participating in the invasion of Yugoslavia and Greece. (NARA)

Sonnenblume – the dispatch of German forces to North Africa

Arrival into North Africa. A Pz.Kpfw. I Ausf. A is lowered onto the docks at Tripoli in early 1941. Due to the hasty nature of their departure, the first German tanks unloaded in Libya still retained their European *Dunkelgrau* colour scheme. No special lubricants were available. Tanks required a maintenance overhaul after 1,000–1,500km – the distance from Tripoli to Tobruk – and an engine replacement after 3,500km (compared to 8,000km under European conditions). While Pz.Kpfw. I Ausf. A in North Africa retained their original engine cooling system, the 'tropicalized' Ausf. B was modified with improved engine air flow (holes cut in the rear deck and engine hatches), larger radiators and a higher fan speed. (Galland Books)

Unloading a kl.Pz.Bef.Wg. in Libya. These vehicles received the same tropicalization as the Pz.Kpfw. I. Note the NKAV smoke-grenade rack.

An early Pz.Kpfw. I Ausf. A pressed into service as a tug for a Junkers Ju87 *Stuka* R-2 (abbreviated from *Sturzkampfflugzeug* or dive-bomber).

As part of Operation *Sonnenblume* ('Sunflower'), the 5th Panzer Regiment was dispatched to Libya in March 1941 with twenty-five Pz.Kpfw. I Ausf. A. A further twenty-five machines followed as replacements, with eleven Pz.Kpfw. I Ausf. A assigned to the 15th Panzer Division. Note the multiple 20-litre *Wehrmacht-Einheitskanister* carried on each vehicle, the white cross denoting water inside.

Rommel's newly arrived armour: a diminutive Pz.Kpfw. I Ausf. A and much heavier Pz.Kpfw. IV during the parade through Tripoli on 12 March 1941.

Approaching Tobruk's outer defences. Although phased out as a front-line vehicle, the Pz.Kpfw. I initially comprised a quarter of Rommel's armoured strength. A subsequent report was quick to condemn the tank as too slow, too weakly armoured and totally unsuited to desert operations.

Pz.Kpfw. I Ausf. A knocked out during the Easter battles at Tobruk in April 1941. German intelligence officer Hans-Otto Behrendt concluded that the first failed attack on Tobruk by the 8th Machine-Gun Battalion and units of the 5th Panzer Regiment was due to 'complete ignorance on the German side of the British [perimeter] defensive positions'. (AWM 007468)

Examining knocked out Pz.Kpfw. I, Ausf. A (top), Ausf. B (bottom).

A British M3 Grant medium tank passes a diminutive, knocked-out Pz.Kpfw. I Ausf. B *Tropen* (tropical). The American tank was used in combat for the first time in North Africa, and with its 75mm, hull-mounted gun it was an unwelcome surprise for the Germans. (AWM 013020)

An abandoned Pz.Kpfw. I Ausf. A from the 21st Panzer DIvision in the Western Desert. Recounting the problems of desert warfare, General Siegfried Westphal wrote of 'a sandstorm followed by torrents of rain, turned the wadis into a sticky morass, so that the troops became hopelessly bogged down in the night, and, in addition, lost all sense of direction'.

Allied investigators noted that the extended chassis of the Pz.Kpfw. I Ausf. B retained '5 bogey wheels, but all of equal size. The forward one was independently sprung on a coil spring with a Luvax shock absorber and the remaining 4 grouped in two articulated pairs mounted on quarter-elliptic leaf springs and connected by a girder. The pairs are arranged so that the tail of the leaf springs points towards the middle. A rear idler was provided, mounted on a cranked arm, on which track adjustment is made.' An additional return roller was also added. It was also noted that an 'interesting characteristic of [all] German tanks is that they all have a front sprocket drive, which means that the power has to be taken through the length of the hull to the gear box and steering mechanism situated in the driver's compartment. There are various arguments for and against the use of a front driving sprocket, but the most common reasons the Germans put forward on its behalf are that (a) it makes the tank more manoeuvrable, (b) it is better to pull a track than push one, (c) since the front of the tank must be heavily armoured in any case, one may as well put the most important transmission units there and so have them protected.'

British recognition chart asking whether the reader could identify a Pz.Kpfw.I.

After inspecting this captured Pz.Kpfw.I Ausf.B at Farnborough, British examiners noted several vulnerable points (identified earlier in Spain): 'The turret ring joint is particularly vulnerable, there being no deflectors fitted for protection. The roller gun mantlet might possibly be jammed by machine-gun or anti-tank fire.'

A US report noted: 'The turret, from which the gunner's seat is suspended, is hand traversed through 360°. The turret floor does not revolve. The inside diameter of the turret ring is 36½ inches. The hull and superstructure are essentially the same as Model A. The suspension differs from Model A in that an additional bogie wheel has been utilized. The trailing idler has been replaced by a rear elevated idler. There is also an additional return roller. The quarter elliptic spring is anchored to the underside of the transverse bogie casting allowing the spring to bump its own opposite end. The coil spring is not used except for the forward independently mounted bogie wheel. There are 99 links in the tracks. The armament consists of two turret-mounted 7.92mm MG 13s. Five smoke candles are carried on a rack at the rear and may be released from inside the tank.' (NARA)

Second series Panzerjäger I in North Africa. An Allied intelligence report stated: 'The fitting of the gun to the chassis is crude in many respects and not up to the usual standard of German workmanship or design. The gun is mounted on a steel frame, within the shield, in the front half of the vehicle. It is a single shot, high velocity weapon. The barrel is a one-piece forging fitted with a large and heavy muzzle brake and flash eliminator. The breech mechanism is of the vertical sliding block type. …

… The gun is automatically cocked when the breech is opened. The elevating gear is on the left-hand side of the gun and is controlled by an elevating wheel with a folding handle. The traversing gear is behind the elevating hand wheel. Traverse is limited by a spring-loaded stop to 15 deg. left or right. The recoil mechanism, which consists of a spring recuperator and liquid buffer, is housed in a cylindrical casing above the piece.' (Galland Books)

Captured second series Panzerjäger I, one of twenty-seven such vehicles equipping Panzerjäger Abteilung 605 in North Africa. An earlier after-action report from France (1940) noted that ammunition stowage boxes would swell 'so that after a few days the rounds don't fit correctly. Then the boxes won't shut.' Experience also showed that the chassis was overloaded by the weight of the gun and surrounding armour. (BA 1011-782-0041-31)

Practical experience in France prompted one commander to report that 'different ammunition types on a Panzerjäger can easily lead to the wrong round being grabbed in the heat of combat'. US intelligence noted the use of armour-piercing (A.P.) and high-explosive (H.E.) shells: 'The gun utilizes the following types of ammunition – (1) A.P. tracer shell, (2) H.E. shell. At 300 yards the A.P. projectile effects a penetration of homogeneous armour of 2.3 ins. At 30 deg. Obliquity; 3.0 ins. Normal, and at 1,000 yards will penetrate 1.8 in. at 30 deg. Obliquity; 2.4 ins. Normal.'

Space inside the Panzerjäger I was limited for rations and personal items, leading to sacks of supplies being carried externally on the track guards. Note the open vision ports, unchanged from the Pz.Kpfw. I. (Galland Books)

Explosive charges demolish a captured *Panzerjäger* I. (AWM)

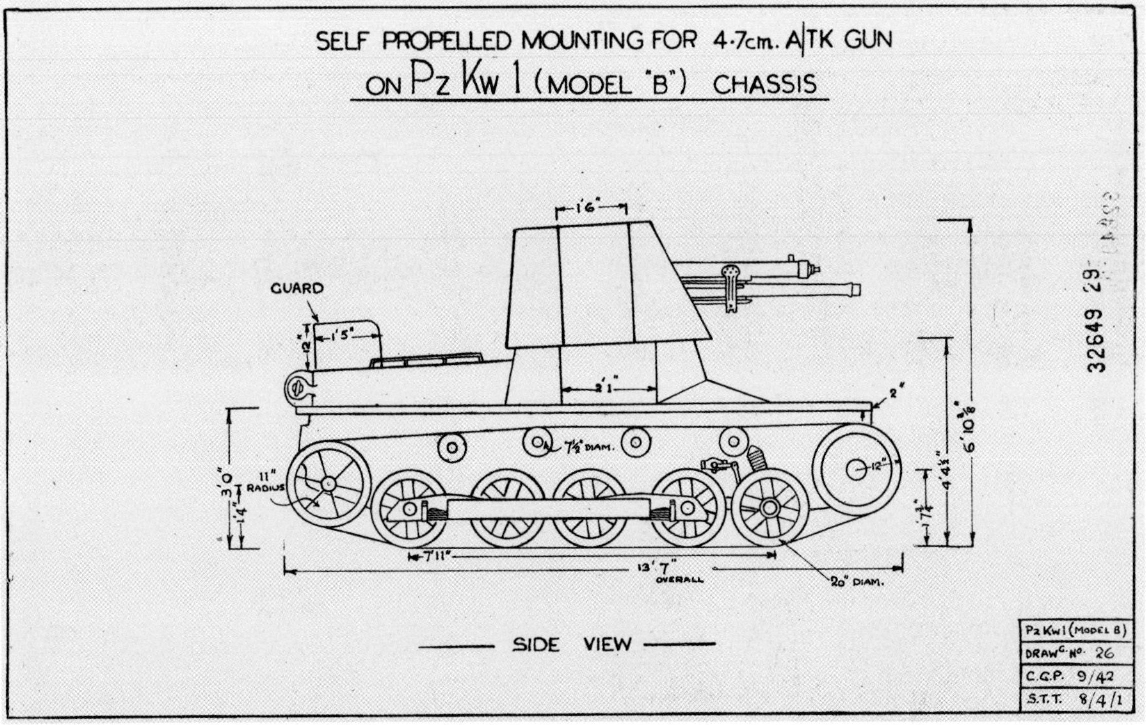

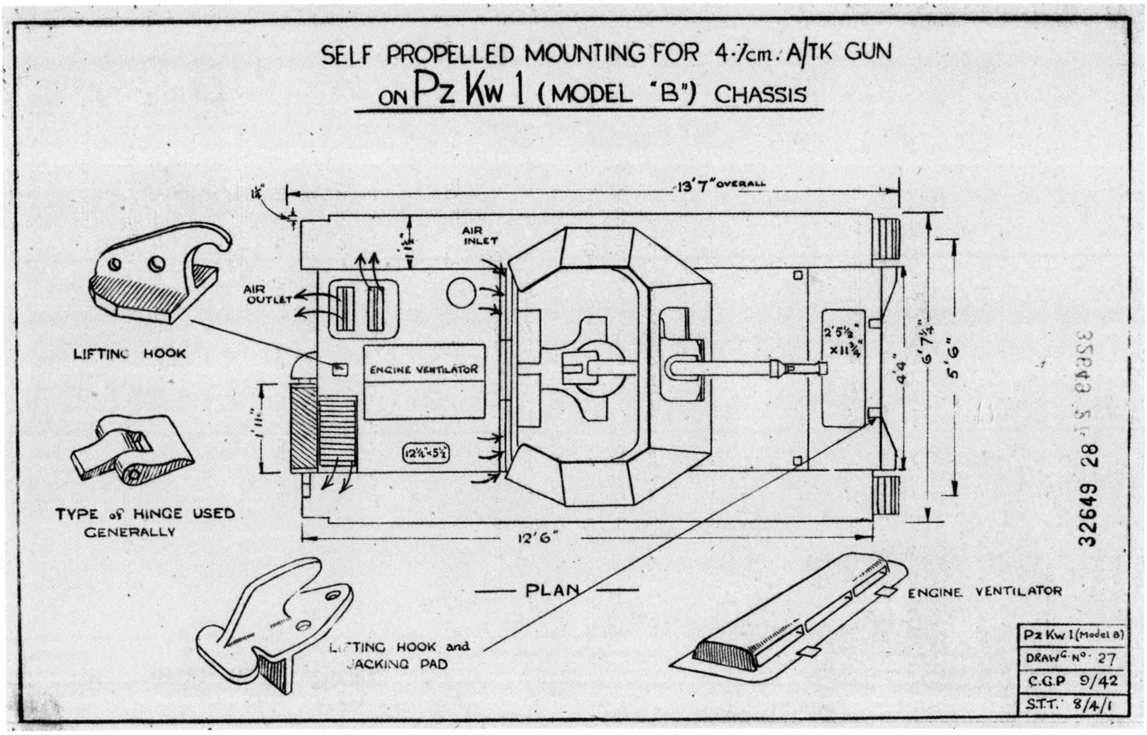

Schematics of a captured Panzerjäger 1 from a British report published in September 1942. Examiners noted: 'As the turret is open at the top and rear the crew is completely vulnerable to attack from these directions. When the gun is elevated or despressed the various sliding plates do not afford perfect protection and would be vulnerable to sustained machine gn fire.' (NARA)

Burning German armour following the capture of a field workshop in Libya in January 1942. All vehicles were destroyed to avert any possibility of returning to service. Note the engine-less *Panzerbefehlswagen* in the foreground (top). (AWM 011030, 011037 & 011038)

Retrieving a knocked out Pz.Kpfw. I Ausf. B. (Collection Akira Takaguchi)

A *kleiner Panzerbefehlswagen* belonging the regimental staff ('R') of the 8th Panzer Regiment (note the insignia on the superstructure side).

NEAR SIDE VIEW

FRONT VIEW

REAR VIEW

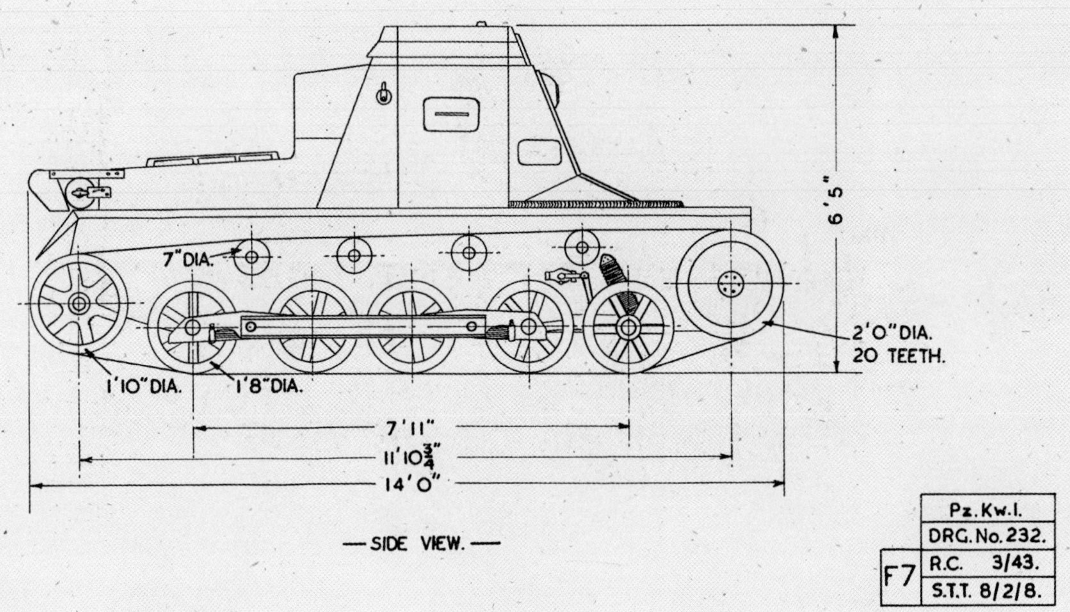

(**Above**) A US report noted: 'This vehicle was planned as a commander's armored office. Many of both Models A and B were used as Commander's tanks. Some were used in the early battles on the Russian front. The chassis is that of the standard Pz.Kpfw. I, Model B, and on this is mounted a fixed square turret with one machine gun in a ball mounting fitted forward to the right. The turret is of octagonal form and comprises one unit with the upper glacis plate, the whole unit being of welded construction. It does not rotate and is bolted to the hull by means of an internal flange at the sides and rear, and by countersunk bolts at the front. A resilient gasket is fitted between the hull and turret. The cupola conforms to the shape of the turret top and is also of welded construction. Its corner plates are short and its internal measurement approximates 20¾ inches. Double slitted visors measuring 8½ ins. × ¾ in. are provided in each side. Entry into the fighting compartment is by a pair of hinged 10 in. × 20 in. doors in the left side plate of the turret. An escape hatch, measuring 20½ ins. square, is provided in the cupola top and is fitted with a pair of doors hinged to the right and left. The hull is of welded construction except where additional armor is fitted, as in the case of the nose plates, where conical-headed bolts have been used. In some instances the extra plates are face hardened. A mounting for a wireless aerial is provided on the right rear side of the turret. The aerial may be raised or lowered from the inside of the fighting compartment by a lever operating on a shaft coupled to the lead-in tube by insulated (porcelain) coupling. Radio communication forms an important adjunct in connection with the operation of these tanks.'

(**Opposite**) An Allied report on the 'Commander's tank' noted a 'normal model B with a built-up superstructure instead of a turret and armed with only one MG in a ball mounting offset to the right. A square shaped cupola with two vision slits on each side is mounted on the right-hand forward edge of the superstructure. The top of the cupola consists of two half-flaps hinged to the side. Double escape doors are set into the left-hand side of the superstructure; the forward door has a vision slit into it. An additional [*Sehklappen*] vision port is located in the middle of the front built-up super-structure. It is protected by a hinged flap incorporating a vision slit. Similar vision ports are provided in each forward side plate and one in the right-hand side centre plate of the built-up superstructure. The driver has the same vision facilities as on Model B. All vision ports are fitted with removable glass blocks and the flaps are operated by levers.'

Reconditioned by the Australian Army Ordnance Corps, this Panzerjäger I captured in North Africa was on display in Melbourne, Australia, during a wartime Liberty Loan Rally. (AWM 138588)

Barbarossa and beyond

The summer of 1941 and a begoggled Pz.Kpfw. I Ausf. B commander advances into Russia. At this stage of the war there were still 337 operational Pz.Kpfw. I in seventeen Panzer divisions. In his speech to the German people justifying the invasion of the Soviet Union, Hitler in part blamed the 'new increase in Russian troops along the German eastern border. Increasing numbers of tank and parachute divisions threatened the German border. The German army, and the German homeland, know that until a few weeks ago, there was not a single German tank or motorized division on our eastern border.' The new Russian tanks would prove to be an unwelcome surprise for veteran German tank crews.

March 1941, pre-invasion. Pz.Kpfw. I from the 10th Panzer Division in Bulgaria cross a bridge over the Danube to Bulgaria.

Kl.Pz.Bef.Wg. in use as a radio vehicle in a mine-clearing company. Note the periscope above the left hatch cover. (BA 1011-265-0006-28)

As well as a command role, the kl.Pz.Bef.Wg. was used as a remote-control vehicle for the Borgward (Sd.Kfz. 300) *Minenräumwagen* (mine-clearing vehicle). Borgward produced fifty B-1 vehicles, a three-wheeled, tracked vehicle featuring a concrete hull that towed a trailer with three rollers. The subsequent B-II model had a more powerful engine and four road wheels.

June 1941. The tactical sign on this kl.Pz.Bef.Wg. indicates the vehicle belonged to a *Minenräum-Kompanie/Abteilung*, later Panzer-Abteilung (FL). FL stood for *Funklenk* (radio-controlled), *Minenräum* for mine-clearing. The rhomboid symbol represents a Panzer; the circle depicts the rollers pulled by the I *Minenräumwagen*. (BA 1011-265-0007-03)

A second photo of the kl.Pz.Bef.Wg, the frame antenna possibly a field modification. (BA 1011-265-0007-02)

Umsetzfahrzeug from the 3rd Panzer Division, Panzergruppe 2 Guderian (note the 'G' on the front hull) with machine gun barrels covered and an assortment of spare track links, road wheels, *Kanister* and Red Army helmets.

Umsetzfahrzeug (front) from the 3rd Panzer Division with additional track on the glacis plate. The *Nachtmarschgeraet* (night driving device) Notek light on the front left track guard, a product of Nova-Technik GmbH in Munich, provided three light settings to reduce vehicle visibility from the air.

УЯЗВИМЫЕ МЕСТА НЕМЕЦКОГО ЛЕГКОГО ТАНКА Т-1

ДИСТАНЦИИ ПОРАЖАЕМОСТИ

1. ИЗ ОРУДИЙ ЛЮБОГО КАЛИБРА С ЛЮБЫХ ДИСТАНЦИЙ ПРИЦЕЛЬНОГО ОГНЯ.
2. ИЗ ПРОТИВОТАНКОВОГО РУЖЬЯ СО ВСЕХ ДИСТАНЦИЙ ПРИЦЕЛЬНОГО ОГНЯ.
3. ИЗ КРУПНОКАЛИБЕРНОГО ПУЛЕМЕТА ПО БОРТАМ, ЩЕЛЯМ И ВООРУЖЕНИЮ СО ВСЕХ ДИСТАНЦИЙ ПРИЦЕЛЬНОГО ОГНЯ.
4. БУТЫЛКОЙ С ГОРЮЧЕЙ ЖИДКОСТЬЮ С 15—20 м.
5. ПРОТИВОТАНКОВОЙ ГРАНАТОЙ С ДИСТАНЦИИ 15—20 м.

УСЛОВНЫЕ ОБОЗНАЧЕНИЯ

Бросай бутылку с горючей жидкостью по щелям и жалюзи.

Стрелково-пулеметным огнем — по смотровым приборам и щелям.

Бей из пушки, противотанкового ружья и противотанковой гранатой по бортам, башне, бензобаку и двигателю.

Г84772

ВОЕНИЗДАТ НКО СССР
Москва — 1942

Типография газеты «Правда». Заказ 1918

Soviet guide to knocking out a Pz.Kpfw. I Ausf. B, highlighting weak spots and aiming points.

Pz.Kpfw. I Ausf. A camouflaged under hay.

Pz.Kpfw. I (likely an Ausf. A – see helmets on turret lifting lugs), Opel *Blitz* 3 tonner and Sd.Kfz. 7 half-track negotiate Russia's *rasputitsa* (season of bad roads).

The 15cm s.I.G. 33 L/11 gun, with an elevation of −4° to +75°, traverse of 5.5° left or right, fired a 38kg high explosive round with a maximum range of 4.7km. Using two-part ammunition, the rate of fire was 2–3 rounds a minute. The cramped vehicle had a crew of four or five. It was not usually equipped with a radio, but this vehicle appears to have one. (BA 1011-216-0406-37)

A motorized column with a 15cm s.I.G. 33 (Sf) auf Pz.Kpfw. I Ausf. B pair during the opening days of Operation Barbarossa, June 1941. A report by schwere *Infanteriegeschütz-Kompanie (mot.S)* 706 noted that the vehicle could not keep pace with an armoured division, advising that it be attached to an infantry division prior to combat. (BA 1011-186-0184-02A)

Umsetzfahrzeug. Caution was always advised when approaching a village. (Galland Books)

Waffen SS *Panzerjäger* I in action. The 6.4-ton vehicle carried eighty-six rounds for the 4.7cm Pak(t) gun, which had an elevation field of –8° to +12°, traverse of 17.5° left or right. Capable of destroying heavy French tanks at ranges up to 600m, the vehicle was also used against fortified positions in urban combat.

German *Gebirgsjäger* shelter behind a Pz.Kpfw. I Ausf. B along the Arctic Front in May 1942, a difficult situation since the tank would inevitably draw enemy fire. Supporting infantry are also at risk of unforeseen, possibly dangerous, movements by the tank. (SA-kuva)

Abwurfvorrichtung auf Pz.Kpfw. I Ausf. B from the 38th Panzer Pioneer Battalion, 2nd Panzer Division, towing a towing a Mercedes-Benz LG 3000 belonging to a repair unit (J = *Instandsetzung*). These specialist tanks were colloquially referred to as *Ladungsleger* (charge layer). (Akira Takaguchi Collection)

Notwithstanding its inadequate armour, the kl.Pz.Bef.Wg. soldiered on in Russia. Entry to the vehicle for the driver and radio operator was via the hatch on the left side of the vehicle; the commander entered through the cupola hatch. (BA 1011-265-0006-16)

A miscellany of light Panzers belonging to the 27th Panzer Regiment, 19th Panzer Division (left to right): Pz.Kpfw. 38(t); Pz.Kpfw. II; and Pz.Kpfw. I Ausf. A. Note the air recognition flag and *Wehrmacht-Einheitskanister*. (Collection Akira Takaguchi)

Red Army soldiers inspect an abandoned 15cm s.I.G. 33 (Sf) auf Pz.Kpfw. I Ausf. B.

A heavily camouflaged 15cm s.I.G. 33 (Sf) auf Pz.Kpfw. I Ausf. B in Russia, June 1942. A handful of these vehicles were still operational in the 5th Panzer Division a year later. It was reported that in emergency situations the gun could be used in an anti-tank role, either deterring oncoming tanks by the effect of the high-explosive shells or striking a tank at a range of 300–400m. (BA 1011-216-0406-32)

Back in Western Europe, African American troops pose for a photo beside a Pz.Kpfw. I Ausf. A in Normandy, 1944, the obsolete tank likely relegated to a task such as airfield defence.

Mock training tank based on a *Fahrschulwagen* I Ausf. A powered by a *Holzgasgenerator* (wood gas generator) in Berlin, May 1945.

Born from desperation, this turretless Pz.Kpfw. I Ausf. B was mated with a 7.5cm KwK L/48 gun (from a *Sturmgeschütz* III or Pz.Kpfw. IV) in the final days of the Third Reich. It is pictured in Berlin's Tiergarten in 1945.

Chapter Eight

Preserved Examples

Pz.Kpfw. I Ausf. A at the Deutsches Panzermuseum, Münster, Germany. (Baku13)

Pz.Kpfw. I Ausf. A displayed at the Arsenalen military vehicle museum outside Strängnäs in Sweden. (Adrian R. Johansson)

Pz.Kpfw. I, seemingly an *Umsetzfahrzeug*, at Patriot Park, Russia.

Restored and running Pz.Kpfw. I Ausf. B at the Australian Armour & Artillery Museum in Cairns.